ON HER MAJESTY'S NUCLEAR SERVICE

COMMODORE ERIC THOMPSON MBE
ROYAL NAVY

CASEMATE

Oxford & Philadelphia

First published in the United States of America and Great Britain in 2018.
Reprinted as a paperback in 2020 by
CASEMATE PUBLISHERS
The Old Music Hall, 106–108 Cowley Road, Oxford OX4 1JE, UK
and
1950 Lawrence Road, Havertown, PA 19083, USA

Paperback Edition: ISBN 978-1-61200-894-3
Digital Edition: ISBN 978-1-61200-572-0 (epub)

A CIP record for this book is available from the Library of Congress and the British Library

Printed and bound in the United Kingdom by TJ International

For a complete list of Casemate titles, please contact:

CASEMATE PUBLISHERS (UK)
Telephone (01865) 241249
Fax (01865) 794449
Email: casemate-uk@casematepublishers.co.uk
www.casematepublishers.co.uk

CASEMATE PUBLISHERS (US)
Telephone (610) 853-9131
Fax (610) 853-9146
Email: casemate@casematepublishers.com
www.casematepublishers.com

We hope you enjoy this book. Please return or renew it by the due date.

You can renew it at www.norfolk.gov.uk/libraries or by using our free library app.

Otherwise you can phone 0344 800 8020 - please have your library card and PIN ready.

You can sign up for email reminders too.

Contents

Acknowledgements

In writing this book, I have been helped, encouraged and inspired by a number of people. In particular, Mrs Margot Aked of Strathkelvin Writers who has voluntarily acted as my mentor/editor over the last two years; my friends in the Helensburgh Writers' Workshop for their critiques since the beginning; and the late Mrs Catherine Lockie who, at the age of one hundred, was still offering me wonderful advice with her characteristic *joie-de-vivre* (RIP December 2016). I also wish to acknowledge the role of the Scottish Association of Writers through whom I have been able to derive the erudite critiques of Stuart Kelly, Literary Editor of the Scotsman, and Professor Bashabi Frazer of Napier University. I must also acknowledge the help and encouragement given by my friend, Madame Dany Lasbleis, whose superb command of the English language has enabled her to vet the manuscript for taste, grammar and punctuation.

For authenticating the Naval content, I am most grateful to my great friend and colleague, Commander Henry Buchanan for checking the entire manuscript; Commander Jamie Dobson for validating the sections on nuclear repair work; Commander Bob Seaward for his advice on the Deterrent; Admiral Paul Thomas and Commander John Osborne for authenticating the *Revenge* incident; Captain Norman Jones, Commodore Barrie Sadler, Superintendent (WRNS) Pip Duncan and Lieutenant Commander Conrad Jones for their help with the torpedo trials sections; Lieutenant Commander John Holl for corroborating the tales from *Osiris*; Commander Tim Cannon for help with the *Warspite* fire and Captain Chris Wreford-Brown for his comments on sinking the *Belgrano*. I have also been greatly moved by the warm-hearted support of my former Commanding Officers, Admirals all, Sir 'Toby' Frere, Sir Christopher Morgan, Frank Grenier, Jeremy Larken, Mike Harris, Paul Hoddinott, Richard Irwin, Tom Blackburn and John Trewby and from Captain Alistair Milne-Home. My thanks also go to Admiral of the Fleet the Lord Boyce for favouring me with a Foreword.

With the help of my friend, Commander Jeff Tall, former curator of the Submarine Museum, I have contacted most of the characters described in the book; none have raised any objections to my depiction of them for which I am most grateful; they are all worthy of great respect.

In finalising the book, the help of my literary agent, Ian Drury of Shiel Land and Associates, has been utterly crucial and without the faith of Clare Litt, Ruth Sheppard and Tom Bonnington at Casemate Publishing, my tale would not have seen the light of day; I am most grateful.

Ultimately, I owe everything to my late wife Kate to whom my book is dedicated and without whom I would not have survived thirty-seven years in the Royal Navy.

Eric Thompson

In memory of

Catriona

(Kate)

Foreword

On 17th September 1961 at the age of eighteen, Eric Thompson and I were marched up the hill from Dartmouth town pier to Britannia Royal Naval College along with one hundred and sixty-three other keen young volunteers who were committing their lives to service in the Royal Navy. We were to be General List, full career officers, the crème de la crème of the officer corps, though few of us felt like that with a Gunnery Instructor snapping at our heels. Thirty-seven years later, Eric retired as Commodore in charge of Her Majesty's Naval Base Clyde, better known as 'Faslane', the principal base for our nuclear-powered submarines. I served on for a further five years – we both stayed the full course. We had also both volunteered for submarines and followed parallel but different career paths, Eric being an Engineer and I a Seaman. We were both engaged 'On Her Majesty's Nuclear Service'.

The secrecy of the Submarine Service means that few outsiders know what life was like for us or what kept us busy. Yet we were performing the greatest public service of all, making a hugely significant contribution to the prevention of a third world war. In this, history shows that we succeeded: the Cold war ended peacefully. It is no coincidence that in the first half of the twentieth century there were two horrific world wars but none in the second half; one could argue that the difference was that in the second half, we had a strategic nuclear deterrent.

As a term-mate and fellow submariner, I have been well aware of Eric's exploits, his superb professionalism, his unique brand of humour and his occasionally subversive outbursts – all captured in this book which provides an authentic historical record, both witty and serious, of what it was like to serve in submarines during the Cold War. It is a good read and I endorse it wholeheartedly.

Admiral of the Fleet the Lord Boyce KG GCB OBE DL

Introduction

My story celebrates the fact that I have lived through the second-half of the twentieth century and have never known a world war. Had I lived through the first-half, I would have faced two world wars, the bloodiest in the history of the human race. I am from the luckiest generation and am truly grateful for that, but peace did not happen by accident; I have lived under a nuclear umbrella through the forty-six years of the Cold War.

After the horrors of the Second World War, Churchill said: '*It must never happen again.*' To ensure it did not, the victors equipped themselves with nuclear weapons, weapons so devastating that they were the ultimate deterrent to a third world war. The principle was called Mutually Assured Destruction.

Inspired by the heroes of the Second World War, I joined the Royal Navy in 1961, volunteered for submarines and served *On Her Majesty's Nuclear Service.* My career spanned thirty-seven years and ended as Commodore in charge at Faslane, the operating base for our Strategic Nuclear Deterrent submarines. I was but one of thousands of men engaged in this peacekeeping mission. We were all anonymous, quietly doing our duty and far from the public eye. Heroes were not required but nor were we robots; we also had lives to lead.

It is a true story, written largely from memory and scrap notes. It is as accurate as I can make it but if there are inaccuracies, please forgive me. In a few places, I have changed the names of former colleagues to avoid unnecessary embarrassment but that does not diminish my tale. This is a definitive inside story of a secret world. It is utterly authentic and, I hope, thought provoking, sobering and, on occasion, amusing. It is my statement of gratitude for the peace I have enjoyed.

Commodore Eric Thompson MBE MSc CEng RN DL

CHAPTER I

On Patrol

'Nobody knows where the submarine goes
And nobody gives a damn.'

GRAFFITI IN A FLEET TENDER

June 1978 – HMS *Revenge* on patrol

The sudden roar came as a shock. It sounded like a jumbo jet taking off.

'Steam leak in the TG room!' a voice shrieked over the intercom.

The roar said it all. This was serious. The Turbo-Generator room was directly beneath me.

Frank Hurley and I exchanged glances. 'Whot-da-fock!' he exclaimed.

We were in the tail end of a nuclear submarine, locked-in behind the massive steel doors of the reactor compartment. Our space was filling with steam. I was Senior Engineer and on watch. My moment of truth had come.

I pressed the general alarm three times – *baaaa baaaa baaaa*: 'Steam leak in the TG room,' I screamed over full main broadcast. There were one hundred and forty men for'ard, not least the Captain. They needed to know; this was a whole boat emergency. In the heat of the moment I forgot to cancel full main broadcast. The entire crew would now be entertained by my new soprano voice – strange how panic reacts on the testicles.

I knew the emergency drill by heart: *Shut both Main Steam Stops.* That would shut off all steam to the Engine Room. At a stroke, it would kill the leak. It was no more difficult than switching off the bedroom lights but it would also scram the reactor, the pumping heart of the submarine; the plant would automatically go into Emergency Cooling and there was no recovery from that at sea. We would have lost our power source, be reduced to a dead ship. We would have to surface and signal for a tug. Unthinkable. *Revenge* was a Polaris missile submarine on Strategic Nuclear Deterrent patrol. She was the country's duty guardian. We were the nation's assurance that World War Three would not happen, not on our watch. We were in our top-secret patrol position. Our number one priority was to remain undetected. Surfacing and calling for a tug would mean breaching one of the country's most highly guarded secrets – where we were. It would mean national humiliation. The credibility of our Nuclear Deterrent was at stake.

If I got it wrong now, the political ramifications would be incalculable. Jim Callaghan's Government was riven by anti-nuclear sentiment. Many of his Labour MPs were proud to flaunt CND badges in public, none more so than Michael Foot, the left wing leader-in-waiting; this could be their golden opportunity. If the Deterrent appeared to fail, British nuclear strategy would be holed below the waterline. Britain could lose its place in the UN Security Council. The Americans could end our Special Relationship. These lofty anxieties flashed through my mind as I prepared to be poached alive.

The Main Steam Stops were operated by push buttons behind my head. I hit the starboard button first. Then a split-second thought occurred. There was a fifty-fifty chance I'd got it right first time. 'Which side?' I yelled into the microphone.

'Starboard,' came a strangulated reply, the voice of Leading Mechanic 'Bungy Mack', a twenty-year-old Liverpudlian on watch below.

Thank God I had not hit the port button for we would have lost all power. But the roar had not stopped. Holy shit! The leak was on the boiler side of the stop valve. One massive, nuclear-powered steam generator was discharging its steam into my airspace and could not be stopped. We were in a race against time. The boiler had to be emptied before it killed us.

There were eight of us on watch. Should I order evacuation now while we could still get out and leave the Prime Minister to deal with the politics? If I did, I would be court-martialled and hung out to dry. The submarine nuclear programme had zero tolerance for failure and I would be made a scapegoat. The roaring continued. The smell of wild steam was spreading fast.

'Steady the Buffs,' I called to the watch-keepers in front of me. That brought a smile to the face of 'Flash' Goodall; he normally supplied the deadpan quips like: 'I've just read the Stores manual. There ain't none.'

It is simply astonishing how the human brain accelerates in a crisis. In a high-speed skid, for example, everything seems to happen in slow motion. That's because the brain has switched to survival mode; it is in hyper-drive. I was now multiplexing at the speed of light. Whilst scanning the vast array of dials before me, I was simultaneously preparing my court martial defence, trying to figure out the quickest way to empty a submarine boiler at sea, worrying about the political ramifications, considering evacuation, and wondering whether or not to scribble a farewell note to Kate in pencil on the back of a log sheet. As there was no risk of us sinking, she would get the note. She was my safe haven. She would still be there for me even if my career did hit the rocks, even if the Navy crucified me. I wondered what she was doing. How I wished I could say: 'Love you,' one last time.

Rapid emptying of a nuclear submarine boiler at sea is simply never done. In any case, the valves to do it were in the bowels of the TG room. How could anyone get to them? By now it would be filled with a suffocating blanket of scalding steam. It was hot enough at the best of times, well over one hundred degrees at the top of the ladder. If he hadn't already evacuated, perhaps Bungy Mack could get below the steam cloud and wriggle through the bilges. Perhaps he was trapped down there. I was also responsible for the lives of my watch-keepers.

In the safety of the for'ard bunk spaces, my cabin mate, Lieutenant Commander Paul Thomas, and two off-watch Mechanicians, McDonagh and Murdoch, were reacting to the alarm. They had heard everything over full main broadcast and unknown to me, had donned fire-fighting suits and come back aft to help. They were already through the reactor tunnel and going down through the steam cloud in the hope of stopping the leak.

'Blow the starboard boiler overboard,' I ordered in the blind hope that Bungy could respond.

'Blowing the starboard boiler overboard,' Paul replied, scarcely audible against the roar of the steam, his microphone much closer to it than mine.

Blowing the hot contents of a high-pressure nuclear submarine boiler into cold sea at depth would be a first. We were about to create the world's greatest ever man-made underwater flatulence. The entire whale population of the Northern Hemisphere was about to be gobsmacked. Submarines hundreds of miles away would hear our indiscretion but, with luck, not recognise it as British. I felt for the Captain. His instructions were *to remain undetected* and I was about to break cover without his permission. For him, it would be akin to having laid an ambush only to hear the regimental band strike-up. I was lucky. Paul Hoddinott was one of the finest Commanding Officers in the Royal Navy. He would understand. Lesser men would have been screaming at me over the Command telephone line – exactly what you don't need in a crisis.

I was now stretching the hallowed nuclear safety rules. If cold seawater came flooding back into the empty boiler, it could damage the hot reactor loop, and we had no way of knowing when the boiler would be empty.

The roaring continued. Air temperature and humidity were rising fast; it was as if an army of Chinese laundrymen were working steam irons in the Manoeuvring Room. I called for a report from the Sound Room.

'No farting heard.'

Gloom and despondency! The blow-down wasn't working. The pressure differential wasn't enough. We were too deep. The drain lines were too narrow. The boiler was merely dribbling overboard and while it dribbled, it remained at full pressure. We would be dead before it was empty. The heat intensified. We were running out of time. Should I evacuate now? We could still abandon the machinery spaces but that seemed like cowardice. And before I could, I would have to put the reactor to bed. I would have to turn us into a dead ship.

'Stop blowing down,' I ordered.

What now? I could think of only one other way of emptying a boiler full of steam. It was another obscure dockyard procedure never used at sea. In this, the main engines are bypassed; the steam is fed directly into the condenser. There, it is cooled and turned back into water. It would be like emptying the domestic hot water tank straight into the toilet.

'Open the Starboard Main Steam Stop. Open the Starboard Dump Steam valve.' I roared into the microphone with no idea if anyone could respond. The Main Steam Stops were quick shutting but had to be hand-cranked open. We had no watch-keeper for that.

For what seemed like an eternity, nothing happened. The roaring of the steam continued. The humidity was unbearable. Our time was running out. Then there was a mighty whoosh down the starboard steam range behind my back. Bless their brave hearts; Paul and his merry men had obeyed the order. The effect of opening these two valves was like letting go the neck of an inflated balloon. The boiler had been deflated. The roar of the steam had stopped.

On the panel in front of me, I watched the dial for the starboard reactor loop pressure fall to zero, as if in sympathy with the boiler. It was a silent statement of finality; a de-pressurised reactor loop cannot be re-pressurised at sea. The massive steam release had actually had a refrigerant effect. That I had not expected. We'd lost one of the reactor's two loops. Worse still, in single loop, although a permissible configuration, we were restricted to quarter power. We could launch our missiles if ordered but our maximum speed was severely reduced. We had no reserve of power for any other emergencies like a major flood or avoiding collision and we were a long way from home. For the next eight weeks, we would be walking a tight rope; one machine failure could bring everything tumbling down.

Submarines on Deterrent patrol do not break radio silence. No one knew of our plight. No one would know for another eight weeks.

At the end of the watch I made my way forward. It was difficult to believe what had just happened. This had been for real; it was not a nightmare. We had been within a whisker of breaching our patrol but thanks to the courage and professionalism of the Backafties (the propulsion engineers), we had kept the show on the road. As I staggered through the Reactor Compartment tunnel into the eerie tranquillity of the Missile Compartment, it was as if I had entered another world in which sixteen one-and-a-half-metre diameter vertical tubes, each containing a Polaris intercontinental ballistic missile with a nuclear warhead, stood like silent sentinels.

The Missile Compartment contained the most powerful weaponry imaginable, each one of these warheads capable of destroying a city. If the rocket fuel in any of these missiles lit up, the submarine would be destroyed in seconds, just like the mighty *Hood*, Britain's largest battleship, when a German shell scored a direct hit on her magazine and she literally disappeared in one gigantic explosion along with the 1418 men in her crew (three survived).

The order of the day in the Missile Compartment was serenity. The missiles had to be permanently ready for launch at fifteen minutes notice and that readiness was tested at random on a weekly basis, but for the bulk of their time the task of the missile-men was to preserve inertia. 'Back Aft' everything burned and turned; steam hissed through pipes, pumps coughed and spluttered, turbines whined and mechanics scaled ladders like hyperactive monkeys. I doubted if the handful of watch-keepers in the Missile Compartment had even been aware that on the other side of the Reactor Compartment, all hell had broken loose.

I smiled. On one of the tubes, a missile-man had stuck a large poster of a naked woman in front of which was a cycling machine; I guess that's one way of relieving sexual energy on patrol. Further along, sitting on a swing slung between the first two missile tubes was Lord Charles, a well-known ventriloquist's dummy and a remarkably accurate effigy of 'Scruff' Hewitt, the Polaris Systems Officer.

When I reached my cabin, I gazed into the mirror. I had grown a black bushy beard. The last time I had grown a beard was back in *Otter* and that had been a disastrous time in my career. Had I tempted Providence by growing it again? We had just suffered a major steam leak. We were in a crippled state. The beard had brought bad luck. It had to go. I stood razor in hand, poised to strike. But I did not believe in luck; nuclear submarines do not run on luck. To shave or not to shave, that was the question. I put the razor down but felt a knot in my stomach.

On my desk lay a message. It was Kate's latest familygram. It read: *Car broken down. Left in middle of road. Large bill expected. Love. Kate.* She was wrestling with her own problems. She had two young children, a large Alsatian dog, an unreliable car, a house to manage, and a job. *They also serve who only stand and wait.* She would not consider herself to be standing and waiting.

At the beginning of each patrol, we made declarations on whether or not we wanted to receive bad news. I had opted to receive it on the basis that if anything disastrous happened to Kate or the boys, I wanted to be with them in spirit at the time and not weeks later. On reading that familygram, I changed my mind. I went straight round to the Wireless Office and amended my declaration. I did not want to hear any more bad news about the car.

When the dust had settled, we discovered the cause of the problem. An obscure drain line from the steam range inside the Reactor Compartment was led out into the Turbo-generator Room, a manned compartment, to allow manual operation when the plant was being warmed up in harbour. It had blown.

Our investigation also revealed that young Bungy Mack had managed to crawl under the steam cloud and wriggle his way through the bilges in a vain attempt to isolate the leak. He had been within three metres of it before being driven back by the heat. In doing so, he had ripped his back open whilst squeezing through a jungle of pipework behind the starboard turbo-generator. At the time, I had no idea he was attempting this. Had I ordered evacuation, he would not have heard the order and could have perished. On discovering his individual act of heroism, I wrote a citation for a bravery award.

CHAPTER 2

In the Beginning

'Many are called but few are chosen.'

MATHEW 22.14

My father was not so lucky. His ship was sunk. German bombers sank it at Leros in the Dodecanese Islands on 26th September 1943 when I was still in my mother's womb.

Until 1939, my father, a slim, diminutive, intensely private man of unimpeachable integrity and impeccably good manners, had been the very model of a law-abiding British citizen. In his own quiet way, he had been pursuing personal happiness as a Glasgow banker and part-time pianist in a local dance band. He had also been pursuing my mother with amorous intent but in 1939, his modest idyll had been shattered by the spectre of Nazi invasion. Along with millions of other peace-loving young men of his generation, he had committed himself to the defence of freedom and volunteered for service in the Royal Navy.

When *Intrepid* sank, the Greek Resistance spirited him away to the nearby island of Kastelorizo in neutral Turkey where they were hidden in the hold of a Turkish merchant ship. It then rendezvoused with a Free French destroyer which delivered them to Beirut whence they travelled overland to British-held Alexandria in Egypt. From there, a troop ship brought him back to England, a further high-risk voyage as the Mediterranean and Western Approaches were still graveyards for Allied shipping. The rest was down to the London Midland Scottish Railway Company.

Intrepid had a busy war. She had rescued troops trapped on the Dunkirk beaches, laid mines off the German coast, escorted shiploads of tanks up to Murmansk for the Red Army, searched for the mighty German battleship *Bismarck*, sunk a U-boat, participated in the deadly supply convoys to starving Malta, covered the Allied landings in Italy and escorted the surrendered Italian fleet from Taranto to Malta before heading up to her grave in Leros.

The Second World War ended when the first atomic bombs were dropped on Hiroshima and Nagasaki and the never-surrender Japanese surrendered. Mankind had experienced an irreversible change. The world had witnessed the power of nuclear weapons, weapons of such horror that they would become a deterrent to war between nuclear states.

The war barely penetrated my infantile comprehension but three things did stick in my mind: the first was Leros, a mysterious name with profound family significance; the second, that Father had a Captain called Kitcat, which I linked to chocolate biscuits of the same name; the third, that war was the worst thing that could happen.

At home, however, the war did not end until Father was demobbed. I remember the excitement but had no idea of what it meant. I presumed they were going to remove his mobs which for all I knew could have been a surgical operation. More importantly, his triumphant return inspired me to join the Navy.

When the war ended, the world moved seamlessly into the Cold War. To quote Churchill: '*From what I have seen of our Russian friends and allies during the war, I am convinced there is nothing they admire so much as strength, and there is nothing for which they have less respect than weakness, especially military weakness. For that reason, the old doctrine of a balance of power is unsound. We cannot afford, if we can help it, to work on narrow margins, offering temptations to a trial of strength. We must not let it happen again.*'

Not letting it happen again was the challenge for my generation; it meant that Britain had to have nuclear weapons.

****

My ambition was kick-started fourteen years later when Father spotted an advert in the *Glasgow Herald* for scholarships to Britannia Royal Navy College, Dartmouth, the college to which the Royal Family sent its sons; where Prince Philipos of Greece was introduced to the Princess Elizabeth before her accession to the throne. Folks in Coatbridge, all well versed in such matters, assured me that as I did not know Lord Mountbatten, I had no chance. Undeterred, I applied and was duly called to the Admiralty Interview Board in Gosport.

At Glasgow Central station, I spotted a boy wearing a Glasgow Academy blazer catching the same train. The Academy was one of the top private schools in the city and my sense of social inferiority kicked in immediately. I guessed he was also heading for the Admiralty Interview – confirmed the following morning when I spotted him on the platform at Waterloo. There, a small army of teenage boys was waiting for the appointed Portsmouth train: some being waved off by financially optimistic parents, some in school blazers, others in sports jackets and flannels, most seemed older than I and were talking in posh English accents. My hackles rose.

When the train arrived, I managed to bag a window seat in an empty compartment but one of the other boys came in to join me. I guessed immediately that he was a rival candidate. He was wearing a light-brown sports jacket, light-blue cotton shirt, woven-silk school tie and had khaki twill trousers that looked as if they'd actually been bought for him. On his feet he wore suede ankle-boots known in the Navy as brothel-creepers. I was wearing my father's best grey flannels; my school blazer with badge on the breast pocket; drip-dry white shirt; nylon school tie and well-polished, black leather shoes. His only luggage was a leather Gladstone bag; mine was the family holdall. His tousled blonde hair looked like it had never seen a comb. My short-back-and-sides had been plastered down with Brylcreem, all the rage then for young rock 'n' rollers like me but not, it appeared, for Public School boys. He looked every inch a young lord of the manor. I felt every inch a State schoolboy – and was proud of it.

He stood for a moment, apparently perplexed, and surveyed the compartment. Then, with much huffing and puffing, threw himself down on the seat diagonally opposite on the corridor side and glared across as if I should have offered him my window seat. His Gladstone bag was very pointedly given a seat to itself; my holdall had been condemned to the luggage rack. When he had finally settled, he looked at me again and with imperious self-confidence and plummy English accent, said: 'Wilmott Randall.' There was no 'Hello' or 'How do you do?', just 'Wilmott Randall.'

When I responded by addressing him as Wilmott and introduced myself as Eric, he told me that he was Wilmott-Randall with a hyphen and that in his school, people were addressed by their surnames. When I asked which school that might be, he informed me that he was a Carthusian. Wow! Charterhouse was Baden-Powell's old school – I'd read that in my *Eagle* comic. Instantly I felt like some low-caste, Indian punkah-wallah.

It was clear that Wilmott-Randall was approaching the Admiralty Interview Board with a presumption of success. No doubt, I thought, he would have royal connections and his father would have been at school with Baden-Powell. It felt like my chances of winning a Scholarship had already evaporated and I hadn't even made it to the interview.

As if that were not enough, Charterhouse was a private school that actually played proper football. It had virtually invented the game back in 1862, fifty years before my school was built, and it was a founder member of the Football Association. Old Carthusians were the first winners of the FA Cup (in which they beat Old Etonians). At least we had football in common, or so I thought. But Wilmott-Randall didn't play soccer as he called it. He didn't even play rugger. He shot. Mon Dieu! If I'd taken a gun into Coatbridge High School, I'd have been handed over to the police.

If only I could have got the upper class twit on to a football pitch, I'd have shown him who was top dog. In my bones I just knew that I would be better than Wilmott-Randall at almost anything he cared to name, with the exception of bullshit and shooting. So why on earth did I feel any sense of social inferiority? The grammar schools were winning the battle for meritocracy but were still losing on social status. The British class system was alive and well.

At Portsmouth Harbour Station, an articulated ramp led down to the Gosport ferry pontoon. Around it, emerald-green seawater swirled and frothed like a river as the ebb tide drained from Portsmouth harbour into the Solent. The ferry was waiting alongside and five minutes later we were across the narrows into Gosport. There, two dark-blue Royal Navy buses were waiting for us. Without ceremony, they whisked us off to HMS *Sultan*, a massive Naval shore establishment in the corner of which stood the two-storey building that was the permanent home of the Admiralty Interview Board.

On arrival we were given a briefing, allocated cabins, and reconvened for early supper, as dinner is called in the Navy. After that we were herded back into buses for a visit to a submarine in nearby HMS *Dolphin*, home of the Submarine Service. That was a shock. It was dark. I had no idea where *Dolphin* was, and I was in a submarine.

Things had suddenly got real.

The interview process began first thing in the morning and was thorough to say the least. It began with devilled kidneys on toast for breakfast – a damn sight more exotic than porridge. I don't know if we were being assessed whilst tackling this delicacy but it could have been our first aptitude test. Could we handle officers' food? Did we know how to use a knife and fork? Were our table manners of Wardroom standard? Or perhaps it was merely intended to test our intestinal fortitude in coping with Naval catering. I didn't like kidneys but was not going to show it. I scoffed the lot and just managed to avoid throwing up.

The President of the Interview Board was a venerable Naval Captain, his panel consisting of the headmaster of an English Public School who knew the ropes, a psychologist, and a Commander. Over the next three days we faced individual and collective tests in front of some or all of them plus having two sessions of written intelligence and aptitude tests. I remember little of the interviews except that the interviewers all asked the same obvious question: 'Why do you want to join the Navy?' My answer was always: 'To defend my country.' 'So it's a vocation?' To this I would reply: 'Yes,' thinking it was a damn silly question. Why else would one join the Navy?

The most memorable of the interviews was with the psychologist. When I entered the room, he was sitting at a desk smoking a pipe and remained silent with his back to me for what seemed like an eternity. I took this to be some sort of psychological test, perhaps to test my ability to remain silent in the company of a more senior person. When he finally turned round, he looked me straight in the eye, pointed his pipe at some indeterminate point over my shoulder and asked what I thought of my mother. 'She's about five feet tall, prematurely grey, and has a great sense of humour,' I replied. And that was the end of the interview. It was such an abrupt end, I felt certain I had failed. In retrospect, I guess his role was simply to find the mummy's boys.

On the second evening, we were assembled for a group discussion in front of the whole Board. This involved candidates sitting round a large oval table with a topic being tossed in by the President for debate. Our topic was to choose the most famous woman in history and give our reasons. After a few dismally unimaginative suggestions like 'Good Queen Bess because she defeated the Spanish Armada,' I proposed Helen of Troy 'because her face had launched a thousand ships.' Then, as a throwaway remark, added, 'which was ten times more than the Admiralty had achieved since the end of the war.' At that, the President terminated the discussion. My instinct for flippancy had been noted.

The grand finale of the Interview was a practical test in the gym. There, we were divided into teams and each given the task of being team leader. The challenge was to lead one's team across a crocodile infested chasm, delineated by two rows of gymnasium benches. These had been placed too far apart to jump across but there was a rope dangling just out of reach in mid-chasm. As I was not first to go, I was able to assess the techniques of the others as one by one they made a hash of it. Then it was my turn.

For once I felt confident. Having seen how not to do it, this was my big chance to show how it should be done. My strategy was to use the team's help to retrieve the suspended rope. Then I would swing Tarzan-like across the chasm and drop down on to the far bank. From there, I would be physically out in front and best placed to shout instructions to the others. Mine would be a display of how to lead from the front and establish a strategic command position – just what the Board would be looking for.

I got off to a promising start. The rope was hooked and pulled back to our side. I grabbed it, took a run at the chasm and swung out boldly towards the far bank only to discover that at the extremity of the outward swing, I was not actually above the opposite bank. If I let go, I would be dropping into the jaws of the crocodiles. The natural inefficiency of the pendulum was then revealed; at the end of my return swing, the rope failed to reach its starting point. If I let go there, I would also have fallen amongst the crocodiles. And so I was condemned to swing gently backwards and forwards with ever decreasing amplitude until I came to a halt, hanging vertically in mid-chasm. There I had to hang-on for dear life until my time was up, or fall to my death amongst the crocodiles. Disaster. I had failed the test.

On the third day, we were packed off to Queen Anne's Mansions in London for a full medical examination. This was no ordinary health check. This was to ascertain whether or not we would be fit to pilot supersonic jets from the heaving decks of aircraft carriers, conduct daring submarine operations at night or lead a team of underwater saboteurs like the Cockleshell Heroes. The problem for me was the eye test; in particular the identification of red, green, or white lights that were presented two at a time in the dark with ever diminishing intensity. It was clearly vital that a Naval Officer could identify port and starboard navigation lights at night so I had to pass this test. When I could no longer see the lights, I decided to blind-guess. At three-to-two against, the odds seemed pretty reasonable. And that was it, Admiralty Interview complete. Parents of unsuccessful candidates were to be informed by letter within forty-eight hours.

Two days later, a brown On Her Majesty's Service envelope addressed to my parents, tumbled through our letterbox. The first line of the letter read: *Their Lordships regret to inform you that your son has failed...* My heart hit the floor. It was game over. Hopes dashed. But no, the sentence continued to say that I had failed only on colour-blindness and that a scholarship and cadetship at Dartmouth had been reserved for me pending successful re-examination of my eyesight. Glory Halleluiah! I knew I wasn't colour-blind.

The subsequent eye test confirmed that I was merely short-sighted and would not qualify for flying or for being in command of a warship but was eligible to join the Engineering or Supply and Secretariat specialisations. Not what I wanted but at least I had been accepted. As I didn't fancy being a secretary, I took a punt with Engineering about which I knew nothing. I also opted for the Electrical specialisation which promised sexier subjects like radar and telecommunications. The catch was that the Engineering Branch demanded the highest academic entry standards of all. I was within reach of my target but now had to pass Advanced Mathematics in my Sixth Year.

The outcome of the Interview had been beyond my control but passing exams was not. It was now down to me. *Laborare est Orare* (To work is to pray) was the school motto. I had never quite grasped its meaning but would now have to both work and pray.

CHAPTER 3

The Four-Minute Warning

'Sleep with Brenda'

DAILY MIRROR HEADLINE

While I was growing up, the nuclear arms race was also developing and impoverished post-war Britain, under the Labour Government of Prime Minister Clement Attlee, in strictest secrecy, developed its own nuclear weapons. It had been the first country to begin atomic weapon development and was determined to remain a top world power.

The nature of Britain's nuclear secrecy was revealed in an incident in 1956 when an ex-Navy diver called Commander Lionel 'Buster' Crabb disappeared in Portsmouth harbour whilst on a top-secret spying mission underneath the Soviet cruiser *Ordzonikidze*. The ship had brought the new Soviet leaders, Khrushchev and Bulganin, on the first ever visit by Soviet leaders to the United Kingdom. It was a State Visit of massive diplomatic importance, described in Parliament by the then Conservative Prime Minister, Anthony Eden, as 'the beginning of the beginning' of a thaw in Cold War relationships. Alas, Crabb's spying mission went badly wrong. The Soviets spotted him in the water and he was never seen again.

In Parliament, the Prime Minister sought to draw a veil over the affair by stating that: '*it would not be in the public interest to disclose the circumstances in which Commander Crabb is presumed to have met his death...what was done was done without the authority or the knowledge of Her Majesty's Ministers.*' Mr Gaitskell, Leader of the Opposition, did not agree. He sniffed a cover-up and accused the Government of failing to control the Secret Service. In defending the cloak of secrecy, the Prime Minister replied that the Government often had to hide things from Parliament, citing: '*the classic example of the atomic bomb, where the whole expenditure was concealed.*'

Four years after the war ended, with the help of atomic secrets leaked by Communist sympathisers in the Manhattan Project, the Soviet Union shocked the West by detonating its first atomic bomb. Three years later, Britain followed suit. The following year, the Soviets and Americans announced that they had developed hydrogen bombs; thermo-nuclear weapons that were substantially more powerful than the crude atomic bombs used against Japan and, more significantly, small enough to fit into missile warheads. On the Pacific island of New Malden, Britain also detonated its own hydrogen bomb to become the world's third thermonuclear power only four years after the Soviets and Americans.

The most significant outcome of Britain's nuclear success, however, was political rather than military. It restored Britain's crumbling status as a world power and prompted the US-UK Mutual Defence Agreement, the Special Relationship, which allowed full exchange of nuclear information between Britain and the United States. Ever since, the two nations have remained nuclear partners.

By the time I was fifteen, the United States and the Soviet Union had developed Intercontinental Ballistic Missiles (ICBMs) tipped with hydrogen bombs, and missile early warning systems were being put in place.

In response to Soviet nuclear advances, the North Atlantic Treaty Organisation (NATO) was formed; its founding principle being that an attack on one was an attack on all. Thus, if any NATO country bordering on the Soviet Union were to be invaded, we would be plunged into nuclear war. In turn, the Soviet Union established the Warsaw Pact which embraced all of the countries that had fallen into its sphere of influence at the end of the war. All military forces on either side of the Iron Curtain were now consolidated into two enormous, nuclear-weapon-armed military power blocks with East and West Germany being on opposite sides.

But there was one essential difference; NATO members were volunteers and Warsaw Pact members were not. When Hungarian President Imry Nagy informed Moscow that his country would be leaving the Warsaw Pact, the Red Army rolled into town. Thousands of Hungarian civilians were killed on their own streets and a quarter of a million had to flee their homeland. President Nagy and a thousand of his supporters were immediately deported to Russia where over three hundred of them, including the President, were executed. All of this happened only months after the Red Army had shot dead fifty-three civilians in the streets of Poznan in Poland. To ensure that no other Warsaw Pact country got similar ideas, embryonic freedom movements in Czechoslovakia and Romania were ruthlessly snuffed out. This was the iron fist of Communism in action. It was no better than Fascism. It was utterly anti-democratic with human rights being subordinated to the State or, more accurately, to the Party.

Such was the alarm in Coatbridge High School that pictures from the Illustrated London News of Soviet tanks in the streets of Budapest were pinned to the school notice boards. I feared that the Third World War would be upon us before I had even joined the Navy.

****

To raise the cash for a top-of-the-range Dawes Double Blue racing bike that Father said we could not afford, I began a newspaper delivery round and one morning noticed the *Daily Mirror* headline: *SLEEP WITH BRENDA*. It was the winning answer to a reader's quiz on what to do on receipt of the Four-Minute Warning from the recently completed Fylingdales Early Warning System in Yorkshire.

On return to the shop, I broached the subject with Jimmy Stewart, the Head Paperboy. He didn't do a delivery round as he was senior management. His job was to make up the bundles for the delivery boys and keep the counter stocked for the Lowrie-esque working masses that came in for papers and fags en route to the local factories. Without exception, these men wore navy blue overalls, flat caps and steel-tipped boots, and all carried sandwich tins.

'This is crap,' Jimmy observed, waving a copy of the *Financial Times* at me.

Why the *Financial Times* was on sale at Coatdyke Cross defeats me. This was the epitome of a working class community; the *Daily Worker* was far more appropriate.

'I agree,' I replied, matching his wisdom. 'By the way, have you heard of the Fylingdales Early Warning System?'

'Naw,' he grunted, 'Ah'm no interested.'

'It'll give us four minutes warning that we're about to be melted by Russian missiles.'

'Complete waste of time,' he growled. 'It would take them more than four minutes to find me.'

It was a fair point. How could everyone in the country be sure to receive the warning and take shelter within four minutes?

'But what would you do if you did get the Four Minute Warning?' I persisted, more concerned with principle than practice.

'Go down the arcade and try for a last jackpot.'

'What's the point of winning a jackpot when you're about to be vaporised?'

He ignored the question.

At that moment, Old Man Crabtree, a jaundice-complexioned Yorkshireman, appeared Fagan-like through the curtain over the door into the back shop, his thinning brown hair plastered down with Brylcreem.

'Mr Crabtree, what would you do if you had only four minutes left to live?' I asked.

He removed the Sherlock Holmes pipe from his mouth and banged it on an ashtray. 'What would I do in my last four minutes on God's Earth?' he asked rhetorically, repeating my question. 'Lad, you know that's a right good question. If you always bear it in mind, you will always be clear about your priorities in life.'

What would I do if I had only four minutes left to live? When on watch in nuclear submarines, I would often pose myself that question: what would I do if I knew the submarine was going to sink in four minutes time but my fairy godmother would grant me one final wish? My answer was always the same: to be in bed with my wife and children, cat and dog, enjoying the blessings of love till death did us part. It's what we used to do on Sunday mornings when my family was young; what I called 'paradise on earth.' The real question is: if you are clear about what you would do in your last four minutes, why not do it now when there's plenty of time?

Whilst engaged in my paper round, my body was engaged in puberty. It was a bad time to choose. Had I adolesced in the Depressing Thirties, I could have been a war hero. Had I waited for the Swinging Sixties, the girls would have been on the pill and I could have taken up sex as a hobby. But I had been allotted the Fearsome Fifties when sex was in the doldrums and nuclear obliteration occupied our minds.

Not everyone rejoiced in nuclear weapon development. In 1957, the Campaign for Nuclear Disarmament (CND) was formed, its first president being the celebrated philosopher Bertrand Russell. Thereafter, CND became the rallying point for a mixed bag of pacifists, intellectuals, dissenting scientists, Christians, and the Left Wing of the Labour Party amongst many others. Anti-nuclear demonstrations then became frequent, popular, and socially acceptable events but achieved little more than keeping the horror of nuclear weapons in the public eye. The Government, however, considered CND to be a potential threat to national security on the grounds that unilateral British nuclear disarmament would serve only the interests of the Soviet Union where no such anti-nuclear protest group would have been tolerated.

****

In 1954, the United States Navy launched the USS *Nautilus*, the world's first nuclear-powered submarine. By the time of my Admiralty interview, she had logged her first sixty thousand nautical miles underwater, matching Jules Verne's fictional *Nautilus* in his classic novel *Twenty Thousand Leagues under the Sea*. Science fiction had become reality.

In the Second World War, Hitler had almost starved Britain into defeat with a flotilla of diesel-driven U-boats, but they had a serious weakness; they had to surface for air to recharge their batteries and could thus be detected by radar and attacked. When airborne radar was invented, they became sitting ducks. The nuclear-powered submarine has no need to surface as it is air independent, like a man-made fish with artificial gills and limitless underwater endurance. It can operate underwater at higher speeds than surface ships and do so regardless of surface weather conditions. It can also circumnavigate the globe non-stop underwater. The nuclear submarine was the greatest underwater development since the fish first wagged its tail.

To emphasize the point, in August 1958, *Nautilus* sensationally sailed under the Arctic ice cap from the Pacific to the Atlantic. Until then, few had imagined that travelling under the polar ice cap was even a possibility – but don't try this at the South Pole.

Six months later, USS *Skate* went one better. She surfaced at the North Pole through a hole in the ice. Pathé News proclaimed that she had: *surfaced where no ship has ever surfaced before*. (Ships do not surface; they sink). Years later, when HMS

Superb emulated *Skate's* achievement, she signalled to Submarine HQ: *Surfaced at North Pole*. To this she received the commendably brief reply: *Steer south.*

If one casts aside the London-centric, imperial British map of the world and looks instead at a globe, the significance of nuclear submarine operations under the Arctic ice cap takes on an entirely different perspective. The Arctic is the meat in the sandwich between North America and Russia; it is the natural battleground for these opposing Cold War Titans, and the warriors in it would be nuclear-powered submarines. Even as a schoolboy, it seemed clear to me that nuclear-powered submarines would change the face of naval warfare.

CHAPTER 4

A New Religion

'God where am I?'

CAPTAIN JACK BROOME, 'MAKE A SIGNAL'

On 16th September 1961, I stood with my parents beneath the smoke-encrusted roof of Glasgow Central station. Platform 1 was on the extreme left, curving gently along the line of the station's high sandstone wall until the track snaked out over the Clyde; I was catching the night sleeper to London. This time, I was off for good. I was seventeen and was joining the Navy. To stave off the lump in my throat, I walked to the far end of the platform and inspected the great steam locomotive that was about to power me through the night to London. Leaving home is one of the great departure points in life.

It had been apparent since winning my scholarship that Father was intensely proud of my going to Dartmouth. Even so, I was taken aback when he whispered in my ear: 'Wish I was coming with you.' Despite the horrors of war and the sinking of his ship, he loved the Navy. Subliminally, he had transfused that love into me.

There is something deeply spiritual about the Navy. I was about to be received as a novice into an ancient brotherhood with its own mysterious rituals and the noblest of traditions. I was now committing myself to the great god Duty: duty to country, duty to ship and shipmates, duty to the defence of freedom and, should the need arise, to the sacrifice of my life, as hundreds of thousands of naval men had done through history. To join the Navy was to enter a new religion. It had martyrs like Nelson, saints in the form of heroes, and its own Holy Trinity – Her Majesty the Queen as the Godhead, the First Sea Lord as its Saviour, and the Minister of Defence as the Unholy Ghost.

At Paddington, it was déjà vu – there was no mistaking the other Dartmouth-bound young men. We were all waiting for the appointed train to Kingswear but this time we were in it together. We were about to become comrades-in-arms.

I had now cast off my school blazer and was wearing a new Harris Tweed sports jacket, cavalry twill trousers, and a Tattersall checked shirt. To signal my urbanity, I had borrowed one of Father's business ties and to preserve my rock 'n' roll credentials, wore bright yellow socks and green suede shoes. It is probably true to say that, as a Dartmouth entrant, I was in a sartorial class of my own. I may have eschewed my kilt and brogues but no one was fooled by my accent. To my complete surprise, I was now being addressed as 'Jock'. Never before had I been called Jock but this was to be my new name. In keeping with joining a new religion, I had been re-baptised.

En route to Kingswear, I treated myself to a full British Rail lunch in the dining car, complete with glass of wine and liveried waiter service – all funded with money my parents had given me for the journey. By the time lunch was over, we were past Exeter and rumbling along the base of the red sandstone cliffs at Dawlish where the waters of the English Channel lap against the edge of the track. These were the waters of Drake and Nelson; 'Royal Navy' seemed to be engraved on every wave. Then we were in Kingswear where the Dartmouth ferry was waiting for us.

During my final year at school, I had read *We Joined the Navy* by John Winton, a hilarious tale of young officer cadets at Dartmouth. (The film version was made while I was there). As I had no other source of knowledge on what lay in store for me, this was my gospel. It was entirely irreverent and taught me to see the funny side of life in a blue suit – *if you can't take a joke, you shouldn't have joined.* John Winton had prepared me perfectly.

The Dart is one of the most atmospheric rivers in England. From the two castles that guard its near secret seaward entrance and between which a chain was once slung to prevent hostile entry, it winds its way through steep-banked Devon hillsides before passing between the twin townships of Dartmouth and Kingswear. From there it meanders up through the mudflats of Dittisham Creek and onwards to Totnes and the high tors of Dartmoor.

The town of Dartmouth has been involved with the Royal Navy since the time of the crusades. It was invaded twice during the Hundred Years War (hence the chain). In 1620, the Pilgrim Fathers berthed there prior to embarking on their epic voyage to America and a life free from religious persecution. Now its harbour was thick with pleasure yachts, small boats, and lines of redundant merchant ships. On a hillside dominating the town stood Britannia Royal Naval College.

The College architect was Sir George Aston Webb whose previous commissions included Admiralty Arch and Buckingham Palace, and Webb did indeed create a building fit for a king, which was just as well. Their Majesties King George V, George VI, Edward VIII, Lord Louis Mountbatten and the Duke of Edinburgh, plus several foreign monarchs, would all be trained in it before Prince Charles, heir to the throne, and his brother Prince Andrew, Duke of York, had their turn. King Edward VII had laid the foundation stone in 1902.

At Kingswear, one hundred and sixty embryonic Naval Officers disembarked from the train and clambered aboard the ferry. Five minutes later, we were at Dartmouth town pier where our freedom ended. There were no buses with Royal Navy emblazoned on their sides this time, just one uniformed Petty Officer wearing the badge of a Gunnery Instructor and he was not there to bid us welcome; he was there to form our rabble into a disciplined squad that was about to be marched through the streets of the town and up the hill to the College in full view of the public. Our bags, we were told, should be left on the pavement for collection. This was it. The moment I had been waiting for. I was in the Navy. I had just become its captive, as much as any prisoner-of-war.

The march was a solemn and silent affair with talking prohibited. We had come in all shapes and sizes and in all manners of attire. Most of us had never before marched in a large squad. In reality, few of us actually achieved the status of marching for we were perpetually changing step to get in step with the man in front who was changing step to get in step with the man in front of him. The net result was that we achieved the corporate locomotion of a centipede.

Awkward we may have been but this squad of marching novices was destined to provide the crème de la crème of the future Navy. It would be for us to fill the central column of the officer corps. Two out of three of us were intended to reach the rank of Commander; one or two were intended to go all the way to the very top. Little did I suspect that from this squad, Mike Boyce would rise to become First Sea Lord and then Chief of the Defence Staff before taking his seat in the House of Lords as Lord Boyce of Pimlico and becoming a Knight of the Garter. Jeremy Blackham would become Vice Chief of the Defence Staff and a Knight of the British Empire. Garth Morrison would take voluntary early retirement and go on to become Chief Scout and a Knight of the Thistle, the top honour in Scotland. Rodney Pattisson MBE would go on to win two Olympic Gold Medals and one Silver for sailing, plus three World Championship Golds in the Flying Dutchman Class. Others would go on to achieve success in an incredibly diverse number of ways, even if only to do their duty with honour. I am not aware of any one of us ever bringing the Navy into disrepute.

From that squad would emerge men who would command aircraft carriers and nuclear submarines, toss atom bombs from low-flying carrier-borne jets, lead underwater saboteurs on cloak-and-dagger operations, manage massive technical projects, become Naval barristers, royal equerries, or be the engineers who would make the modern Navy tick. From the ranks of those who retired early would emerge a circuit judge, a professor of astrophysics, merchant bankers, stockbrokers, an investigative journalist of international repute, and three Anglican vicars, one becoming Dean of Christ Church Cathedral in Oxford. Thank God I had no concept of the calibre of my newfound companions.

My ambition had simply been to join the Navy and I had just achieved it. As we marched through the streets of the town and up the hill to the College, my only thought was: would I be good enough? It was a thought that would remain with me throughout my career. My ambition now was simply not to fail. It's negative, I admit, but nevertheless it is the strongest of motivators. I switched to survival mode.

When we reached the great, white, ship's mast marking the entrance to the College parade ground, we were wheeled right and brought to a halt in front of the ceremonial staircase that led up to the main entrance. Wrapped around the parade ground were The Ramps. These provided vehicular access to the main entrance but were more commonly used by Gunnery Instructors to send Cadets doubling round them, rifles held above their heads, for drill errors during parade training – and sometimes just for breathing.

Writ large across the front facade of the College were the words: *It is on the Navy, under the good providence of God, that our wealth, prosperity and peace do depend,* the opening words of the Naval 'Articles of War'. The sentiment jarred on me. I had joined the Navy to defend my country, not to preserve wealth and prosperity. For an officer-training establishment, I would have much preferred the words of Robert Burns: *The rank is but the guinea stamp. The man's the gold for a' that.* Although built at the same time and also with public money, Britannia Royal Naval College was a very far cry from Coatbridge High School, the latter having been designed only for the brainier offspring of the working class; there was no suggestion of wealth and prosperity in its motto: *To work is to pray.*

The College was a mere fifty-six years old but was steeped in tradition. On entering by the great main door, one arrived in a wide, parquet-floored corridor that ran the entire length of the building from Wardroom at one end to Chapel at the other. Along its walls were photographs of all the previous Cadet terms going back to the College's opening, with one showing a class of thirteen-year-olds in 1914, trunks packed ready to go to war. All the young men in this vast array of photographs had joined the Navy just as I was doing. Most had fought for their country; many had made the supreme sacrifice. I was now walking in the footsteps of heroes.

The worship of God has always been at the core of life for seafarers and the Royal Navy is no exception. From the centre of the main corridor, I could look straight into the nave of the Chapel at the far end. The walls of the Chapel were also covered in memorials, in this case polished brass plates in memory of Cadets who had lost their lives in the service of their country. One was particularly poignant; it marked the loss of an entire College class that had gone out for a week of sea experience in the submarine *Affray* never to return; she sank off the Channel Isles.

What happened next remains a blur in my memory. At some stage, we were roll-called into Divisions; I was to be in Exmouth Division. I was then issued with my kit, which included everything I needed for life in the Navy including pay book and gas mask – at last, I had the gas mask satchel I had coveted for a school bag in Coatdyke Primary. Gas masks had been invented for trench warfare in the First World War to counter the descent into chemical warfare but now were of greater importance for surviving nuclear fall-out. My gas mask would prevent radioactive particles entering my lungs and it was made very clear that I had to have it with me at all times.

The pay book was a green canvas-backed document about the size of a passport and bore a multitude of intimidating instructions, such as (in bold type): *should not be allowed to fall into the hands of the enemy.* That seemed entirely sensible – allowing the enemy to draw one's pay would have been very bad for morale.

On Day Two, we were marched into the large lecture theatre and given our sex education Navy-style, cunningly programmed for maximum initial impact. This involved a screening of that classic of Naval instructional cinematography, *The VD Film*, a black-and-white horror movie of wartime vintage – with a little more subtlety it could have been called *Brief Encounters.* It took us on a voyage of discovery through the world of crabs and into the realms of gonorrhoea – '*It's like pissing broken bottles, Sir*' – then descended into the hell of incurable syphilis which, we were informed, could come in some parts of the world in a cocktail (unfortunate choice of metaphor) with leprosy. The film then showed examples of the collateral damage that could be inflicted in a sexual encounter. It showed pictures of men with hairstyles like my father's wearing the same Admiralty Pattern knee-length underpants with which I had just been issued. With the underpants removed, it revealed that the end of their knobs had either rotted away or been blown off. It showed men with pus-filled bomb craters round their genitals; men with testicles riddled with woodworm; and finally the grotesque spectre of faces disfigured by terminal syphilis. Praise the Lord; HIV and Aids had not yet been invented.

After the film, we were introduced to a strange little package called a condom and instructed to have one with us at all times in the company of women. The instructor didn't actually explain whether it was meant to be swallowed in emergency or used as some form of repellent but being a practical young officer, I sellotaped mine to the inside of my gas mask, which I had also been told to have with me at all times. This Naval sex education had such a profound impact on my impressionable young mind that I had been married for six years before my wife persuaded me it was safe to take my gas mask off in bed.

To complete our sex education, a Surgeon Lieutenant took to the stage and uttered the unforgettable advice: 'Gentlemen, never flatter yourselves. If she'll drop them for you, she'll drop them for anyone.'

The Admiralty's position on sex was absolutely clear. Married candidates were excluded; if a Cadet were to attain the state of holy matrimony during training, he would be expelled. If on completion of training, a young officer were to embrace nuptial bliss before the age of twenty-five, he would be denied Officers' Marriage Allowance and a married quarter. And to ensure that a Cadet's carnal instincts did not get the better of him during initial training, we were confined to College for the first six weeks or until we had passed our Parade Training Exam, whichever came later. Celibacy was clearly Their Lordship's intent; even the tea in the Cadet's mess was laced with bromide.

My new bedfellows came from a wide variety of backgrounds. Fifty per cent were from top notch English Public Schools. The rest of us were either from Grammar Schools or the Upper Yardman scheme, the latter enabling exceptionally talented young sailors to apply for officer conversion. Many of us had fathers who had been or were still in the Navy, some being the sons of Admirals, others having lost their fathers during the war. Richard Wraith's father was a young submarine CO whose boat was sunk at the same time as my father's ship, also in the Mediterranean. He too was still in his mother's womb, but unlike me had sadly never known his father. Lyle Ryder was the son of Commander Robert Ryder VC who in 1942, led the breathtakingly daring Commando raid on the massive *Normandie* dry dock in enemy held St Nazaire, the only Atlantic Coast dock big enough for the mighty German battleship *Tirpitz*, which was then about to deploy into the Atlantic to wreak havoc with our convoys.

Although many of my peers had powerful Naval, family or old-school-tie connections, the Naval training system seemed to be absolutely free of favouritism. Indeed, for those with Admirals or war heroes as fathers, the challenge of living up to family expectations could be a heavy burden and proved too much for some. I was lucky – no one expected anything of me so I could not disappoint.

The over-riding purpose of Dartmouth is to develop character, to take the raw material of new recruits and mould it into young officers fit in all respects to go into the Fleet and take responsibility – and the Navy had a tried and trusted method. First it beat us into pulp through humiliation and exhaustion until every instinct for free expression had been expunged. Then it injected its own DNA into our quivering remains. The process combined the best features of genetic engineering and religious indoctrination and had been forged from centuries of experience. Dartmouth DNA is known as OLQ (Officer-like Quality). A Cadet could fail a Navigation exam and be allowed a re-scrub (re-sit) but one who failed to exhibit OLQ would be thrown out.

To ensure that I was well pulped, I had to undergo one round of boxing against another Cadet. I had never boxed before. I had never even put on boxing gloves. But I was confident that I would be able to dance my way round the ring whilst delivering a barrage of devastating blows to my unfortunate opponent. As Mohammed Ali put it, *to float like a butterfly, sting like a bee.*

Though I was slightly taller than my opponent, I was thin and wiry, while he was short and squat. How I pitied him. I was clearly the athlete and I was going to humiliate him. As soon as the bell struck, I was out of my corner like a greyhound out of a trap, gloves held out in front of me in the classic boxing pose I'd seen in newspapers. Two thirds of the way to his corner, I came within his strike range and he hit me straight on the nose with his first punch. And that was that. For the next three minutes, I simply kept my head down and flailed my arms like a manic windmill. I doubt if I landed a single punch but he certainly landed lots on me. Somehow, I survived the three minutes. Afterwards, my opponent told me that he had been school boxing champion. He would have known immediately that I hadn't a clue.

After pulping, the emphasis switched to leadership potential. I was to be assessed for characteristics such as courage, selflessness, determination, and the ability to set an inspirational example. Such skills can scarcely be taught or even defined with precision: some men are born leaders, some have leadership thrust upon them, others learn to lead, some are dry and scholarly but brilliant planners, others have the knack of being in the right place at the right time, and some are simply blessed with charisma. For whatever reason, a leader must have whatever it takes to make others follow.

During an early lecture on leadership, I proposed humility as a desirable quality only to be told in no uncertain terms that humility had no part in leadership. 'Neither,' I replied, 'has pomposity.' Common Sense, I agree, is a valid parameter but Intelligence is questionable. How can an officer show continuous improvement in his intelligence? It is a biological fact that a man's intelligence peaks in his late teens, along with his sexual prowess. It would be a freak of Nature if an officer of middle age were to show a steady improvement in intelligence.

The ultimate role model for leadership is Nelson. He believed from an early age that he was destined for greatness and missed no opportunity for death or glory. For this, he lost an arm, an eye, and ultimately his life, but he saved the nation from Napoleonic invasion and achieved immortality. Nelson was unflinchingly independent of mind and did not hesitate to disobey orders when he had better ideas, most famously at the Battle of Copenhagen when he ignored his Commander-in-Chief's order to retreat by putting the telescope to his blind eye. He won that battle and could claim honestly that he had not seen his Commander-in-Chief's signal – *putting a blind eye to the telescope* is now immortalised as a figure of speech.

Immortal Nelson may be but he was not short on human frailty. Having been seriously wounded at the Battle of the Nile, he was nursed back to health in Naples by Lady Hamilton, wife of the British Ambassador, and fell madly in love with her despite both being married. After a moment of passion in the cabin of his then flagship, HMS *Foudroyant*, she became pregnant and his misconduct became a public scandal.

The spirit of Nelson is the ethos of the Royal Navy and his qualities were now expected of me – but if I were to disobey an order, I would be court-martialled. And if I were to be caught with a girl in my cabin, let alone having sex with an ambassador's wife, I would be out on my neck.

At Dartmouth, I was taught all the professional naval subjects such as navigation, gunnery, air warfare and seamanship but almost nothing about submarines and nothing at all about nuclear submarines. The Submarine Service was still a small, unfashionable part of the Navy. To all intents and purposes, it was a navy within a navy. Few outsiders knew anything about it and those inside were secretive and somewhat subversive. As our first nuclear-powered submarine, HMS *Dreadnought*, had been launched only the previous year and had not yet entered service, there were no nuclear submariners on the College staff – they had still to be bred. Little did anyone at Dartmouth realise that nuclear submarines were about to become the primary arm of British defence.

Even more significantly, the US Navy had now deployed USS *George Washington*, the world's first ballistic missile firing submarine (SSBN), a squadron of such Polaris boats being based in the Holy Loch, only forty miles from home. Father and I drove down to view their long, black, sinister shapes lying low in the water alongside their depot ship. It seemed utterly incongruous to see them lying at peace in the tranquillity of a Scottish loch knowing that each carried more destructive power than all the bombs dropped in World War 2. Each of these submarines could annihilate sixteen Soviet cities in one salvo.

A more comforting thought was that Polaris submarines had not been built to fight wars; they had been built to prevent them. Their role was simply to convince the Soviets that their own destruction was assured should they ever have the stupidity to attack the West. No longer could a warmongering maniac like Hitler feel safe in a bunker in his capital; one Polaris submarine could have taken out the whole of Berlin with just one of its sixteen missiles.

The unique factor in this new strategic dimension was that these underwater launch pads were invulnerable to pre-emptive elimination. When deployed on patrol their location would be unknown and there would be a force of them continuously on patrol and ready to fire. The Age of Submarine-launched Strategic Nuclear Deterrence had begun.

CHAPTER 5

The Small Matter of a Journal

'Theirs not to reason why;
Theirs but to do or die.'

TENNYSON: CHARGE OF THE LIGHT BRIGADE

It is always stirring to arrive at a jetty and find your ship alongside. Arrival by naval bus at a nondescript jetty somewhere in the vast Devonport dockyard where HMS *Vigilant* lay in dead-ship state was less romantic. I was joining the Dartmouth Training Squadron to discover the harsh realities of life at sea. To record the experience, I had to keep a journal.

<u>Journal:</u> 'One thing is immediately apparent; the ship is being manned by dockyard mateys and sailors. The sailors work hard and take pride in the ship while the dockyard mateys spend their days drinking tea, reading newspapers, playing cards and watching the sailors work. It is no coincidence that Parkinson's Law – *Work expands to fill the time available* – was first observed in a Royal Dockyard. Professor Parkinson was an Admiralty Civil Servant.'

Mentor's comment: *This page is for sketches.*

<u>Journal:</u> 'A British sailor is a man of whom the country can be proud. On the treatment of him:

1. Remember he is a human being living under difficult conditions.
2. Always make time to listen to his suggestions or troubles.
3. Always encourage him but never with empty flattery.
4. If he asks for help, give it unselfishly.
5. If you need his help ask him for advice; let him suggest the solution.
6. Avoid giving orders and treating him like a machine.
7. Never ignore him ashore.
8. Always remember Drake's order when his ship ran aground: '*Let the gentlemen haul and draw with the seamen and the seamen haul and draw with the gentlemen and let us be all of one company.'*

Mentor: *Pay more attention to your writing – you make some silly mistakes.*

What! My youthful analysis was the stuff of leadership manuals. I should have been commended for mature insight into man management. I was ahead of my time.

Journal: 'Went over the side on a painting stage, a plank of wood suspended only by wrapping one turn of rope around either end – a bit scary. Later I had to paint the boot topping (the black bit along the waterline); it was cold, filthy, boring work. The ship was then fuelled, sank a few feet lower in the water and the flotsam and jetsam of the dockyard washed up over the fresh paint. Careful planning would have avoided this.'

Mentor: *You are right but things never go according to plan however much we try. The great thing is to make a plan and never lose your sense of humour.*

Journal: 'Today I was sent into Plymouth to sketch the Hoe, the Mayflower Memorial, the Elizabethan House in the Barbican, and the new Tamar Bridge. During the expedition I was smitten with an acute attack of diarrhoea and spent most of the day searching for public toilets. I could have written a better informed dissertation on the City's public conveniences.'

Mentor: *Have you tried your hand with pen and ink?* (I think this referred to my sketches).

Journal: 'Great excitement. We were towed out into midstream to allow the magnificent HMS *Barrosa* to take the inside berth. She is a Battle Class destroyer converted recently into an ultra modern radar picket, easily recognisable by the huge bedstead radar on top of her foremast.

'Went begging for green paint from other ships as the dockyard store had none. When I asked why we needed it, the ship's painter replied: 'For a cocktail party in Copenhagen.' I had assumed it was for camouflage.

'In the neighbouring basin, HMS *Tartar*, one of the Navy's latest frigates, was easily recognised by her twin funnels, the second being required, I presume, for her gas turbines which need large air intakes and exhausts.'

Mentor: *The second funnel is indeed for the gas turbine. The other reason is atomic fall-out protection. The boiler is enclosed in a sealed box and its associated machines are in the Engine Room where they can be operated automatically. Thus, when steaming through Base Surge or fall-out, only the sealed boiler box is contaminated.*

Welcome to the Nuclear Age! The Navy was deadly serious about fighting on through a nuclear holocaust.

Journal: 'Modern warships have gas-tight citadels into which the crew can retreat whilst passing through radioactive fall-out. The last line of defence is my gas mask. That will save my lungs but what about my testicles? They are left hanging out to fry. Why was I not given a lead loincloth? I haven't had any children yet.

'Scrubbed the decks; a waste of time aggravated by the stupid task of drying them after a rainstorm. I learned nothing. What I need is skill training.'

Mentor: *You may well regard this as time wasted. However, your future career as an officer is to lead men. To do this you must have experienced their way of life.*

Journal: 'Seaboat drill was exercised for the first time. It was a disgrace to naval organisation. Two officers and three Petty Officers were all barking instructions at us, a classic example of too many cooks spoiling the broth.'

Mentor: *As you say, it was your first experience. Give the people concerned their due – they are responsible and were more than a little concerned over the possibility of someone being hurt.* (The seaboat was a twenty-seven foot long (9 m) Montagu whaler weighing over a ton).

Journal: 'Sailed into Copenhagen where Nelson put the telescope to his blind eye.'

Mentor: *Sketch of Copenhagen harbour appropriate here.*

Sketch inserted.

Mentor: *Sketches should be on the left hand page.*

Journal: 'In the evening, the cocktail party for which the green paint had been obtained was held. Deck scrubbed during a thunderstorm.

'The Americans have successfully sent an astronaut (Scott Carpenter) into orbit. He returned safely and his capsule was located by SARAH (Search and Rescue Homing Apparatus), the only British equipment involved.'

Mentor: *Carpenter's circumnavigation of the globe was quite an event. You've treated it as if he went down the road to buy a hot dog made from an English sausage.*

Journal: 'I have a fifteen-hour working day. In the remaining nine I have to sleep, keep a night watch, do personal maintenance and write up my journal. I even have to scribble by torchlight with hammocks swinging overhead. Quantity, not quality, seems to be the requirement.'

Mentor: *I agree and point taken – but you must admit it does sort out the sheep from the goats. It won't get any easier as you go on in the Service. It's tough at the top!*

Journal: Today we celebrated the Queen's official birthday at Kolding in Denmark. As *Vigilant* carries no saluting gun and the four-inch is far too large for firing blanks – it would have shattered all the waterfront windows – our forty-two-gun salute was carried out by firing blanks from a rifle up its barrel. The effect was most impressive but the Danes must have wondered why the British were firing rifles up the barrels of their big guns.'

Mentor: *Still not quite the right idea.*

Journal: 'The night before entering Gothenburg was my turn to prepare and execute a navigational passage. The rules for planning a passage are also good rules for life:

1. Avoid all known dangers.
2. Anticipate unknown dangers.
3. Read the advice in books but use only for guidance.
4. Have answers ready for questions you may face.
5. Allow for being wrong.
6. Always have an escape plan.
7. Remember: The best laid plans of mice and men gang aft agley. (Robert Burns)

Mentor: *Better this week. Try and keep it up.*

Journal: 'In our approach to Gothenburg, a streamlined modern submarine suddenly appeared. I thought she must be Swedish but was overjoyed to discover she was British. It was HMS *Porpoise,* our first post-war-designed submarine. She berthed alongside us.

'As I have been thinking of volunteering for submarines, I invited myself on board and expected to find her all mod con with exciting new developments. There were certainly new developments such as the diesel-electric drive that made her the quietest submarine in the world and aircraft-style joysticks were being used to operate the steering and hydroplanes but there were few advances in accommodation. The crew still lived like sardines.

'On board, I received the warmest of welcomes and sensed an altogether more informal atmosphere. In the Officers Heads there was a notice that read: *Do not throw your dog ends in our urinal. We do not urinate in your ashtrays.'*

Mentor: *Don't write such quotes in your journal. Read the instructions at the front.*

When all else fails, read the instructions. Had I read the instructions, I would have fared much better but I regarded my journal as a waste of time. I had not joined the Navy to be a journalist.

Captain's Half-term Assessment: *Most of your comment is ill informed, inaccurate, and tactless, sometimes to the point of stupidity. Try to write some mature, intelligent comment. Remove the misplaced attempts at humour, the lengthy discussions, and lower deck drips.*

Disaster! Tact and diplomacy were essential Officer-like Qualities. This could mean the end of a very short career.

Six weeks later, I penned the final words in my journal: 'In three fantastic months I have learned more about this ship and the Navy than many would learn in years. We have had precious little free time. We have been treated badly, often thoughtlessly, and been sworn at, ridiculed, overworked, undernourished and overloaded with learning. We have averaged five hours sleep per day and been treated like thirteen-year-olds, not as young men of university entrance standard. Leave granted has been to the same standard as a Girl Guide camp. People have tried to prove we are imbeciles and treated us like unwilling conscripts but we are all volunteers and dead keen. In spite of all this, the last three months have been the best of my life.'

Mentor's final comment: *All you say is very true. It has been a hell of a three months. But the reward and the reason for it is there in your own words, in the very first sentence!! People who have lived a hard, disciplined life, suffered and risen above it, are usually worthwhile. The Service with its customs, traditions, training, and record of being second to none has evolved over many centuries – I think it's right.*

You will probably never again in your career go through quite so much for such a long period. However, at times for shorter periods you will be under greater stress. The reason for this is that you will have more responsibility and the ultimate decisions will be yours. These decisions must often be made when you are desperately tired, thoroughly depressed, and for two pins would go home. It is only by your training and experience that you will, in the end, do the right thing and lead men.

I must not preach but I would commend to you two things which I find helpful:

Firstly – Rudyard Kipling's poem, 'If.'

Secondly – When things are tough, say to yourself: 'It's a great life, if you don't weaken, and then smile.'

I had coped very well with the demands of the Training Squadron but the accursed journal had exposed my instinct for challenging authority: nothing wrong with that – rebels are reformers – but I was only a trainee. Before I left *Vigilant*, my mentor told me that I looked too young to be put in charge of men and should grow a beard and take up smoking a pipe. The missing clues! I had already been told to cut the comedy and avoid sex. The perfect Naval Officer, it seemed, was a hairy, pipe-smoking virgin without a sense of humour. I resolved to grow a beard before I joined submarines.

From *Vigilant*, I returned to College for my Passing-Out Parade at which Her Majesty the Queen inspected me.

****

On 14 August 1962, as a newly promoted Midshipman, I joined 'the magnificent HMS *Barrosa*.' I was now up with the likes of Hornblower. Four days later, a pop group called *The Beatles* played their first gig on Merseyside. They were about to go stratospheric; I was about to go to Singapore. They were at the cutting edge of a new British pop culture that would conquer the world; I was at the tail end of a disappearing British Empire.

During my first term at Dartmouth, the Soviet Union had detonated a fifty-eight megaton hydrogen bomb over Novaya Zemlya in the Arctic, the largest ever man-made explosion, fifty-eight thousand times larger than the atomic bombs dropped on Japan. And, horror of horrors for the Americans, the new Cuban leader, Fidel Castro, had announced that he was a Marxist-Leninist and that Cuba would be embracing International Communism. The Soviets had already begun to deliver shipments of Intermediate Range Ballistic Missiles (IRBM) to Cuba under the guise of agricultural equipment. Within nine days, forty IRBMs had been deployed to nine sites concealed under palm trees. The Soviets could now target the whole of the USA from Cuba.

Soviet President Khrushchev's calculation had been that the young President Kennedy was weak and indecisive and would merely accept the missiles as a *fait accompli* when their presence was discovered. Castro's view was more hawkish: if the Americans were to invade Cuba, he would launch a nuclear strike accepting that Cuba would be destroyed in retaliation – an alarming new Latin American variation on Mutually Assured Destruction. Russia would then retaliate by invading West Berlin and that would trigger nuclear war in Europe.

Neither Russia nor America wanted a Third World War but Kennedy had an election to face and had to be seen to act tough. Khrushchev, on the other hand, had the prestige of International Communism at stake and could not be seen to back down. It was power politics at its most terrifying, a game of political poker with the future of the human race at stake. By the time *Barrosa* had reached Aden, America had given the Soviets an ultimatum: if they did not stop supplying arms to Cuba, America would blockade the island. The Soviets refused to give way and both superpowers went on to a war footing. **We were on the brink of World War Three.**

The Cuba missile crisis was the nearest the world has ever come to nuclear war. Thank God (or Lenin), the Soviets blinked first and backed down. As Krushchev said later: '*They talk about who won and who lost. Human reason won.*' The principle of Mutually Assured Destruction had been validated. Churchill had been spot on when he said: '*There is nothing the Russians admire so much as strength, and there is nothing for which they have less respect than weakness, especially military weakness.*'

****

The Cuba crisis had barely subsided when I received a newspaper cutting from my parents. Britain had just signed the Bahamas Agreement with the United States and would be buying the Polaris weapon system. In terms of British Defence policy, this was revolutionary. It meant that responsibility for the national nuclear deterrent would transfer from the Royal Air Force to the Royal Navy; the Navy would now be making a monumental investment in nuclear-powered, missile-carrying submarines. The Submarine Service was about to be rocket propelled to the top of our Defence priorities.

According to the article, British Polaris submarines would operate from a base in Aden and deploy into the Indian Ocean from where British missiles could target Russia's underbelly. The USSR would thus be surrounded by a ring of missile carrying submarines in the Pacific, Atlantic, Indian, and Arctic oceans as well as the Mediterranean. In the Indian Ocean, our submarines would be out of range of Soviet anti-submarine forces because the Soviets suffered a major geographical handicap; their nearest naval bases were at Murmansk beyond the north tip of Norway, and at Vladivostok beyond the northern tip of Japan. The Soviets desperately needed to find friendly countries in Asia, Africa, and the Mediterranean for naval bases.

****

By coincidence, at the end of my year as a Midshipman, *Barrosa* visited Nagasaki, seventeen years after the dropping of the atomic bomb. The city had been completely rebuilt and was now a bustling modern city. The only evidence of its nuclear experience was on display in the Atomic Bomb Museum and associated Peace Garden with its deeply symbolic Peace Monument. The monument sits at ground zero, one hand pointing to the sky where the airburst detonation took place, the other being held out horizontally to symbolise peace on earth. I sketched it for my journal. As I sat sketching, it was difficult to believe that I was sitting at the epicentre of a nuclear detonation.

****

I had flourished as a Midshipman – even scored 88% for my journal – and at the end of the year, received my Queen's Commission. I was now a fully-fledged Naval Officer, but had not even begun training as an Engineer. My next stop was the Royal Naval Engineering College at Manadon, the Plymouth Monosex Monotech. It was the world's most exclusive gentlemen's club, populated only by full-career, male, commissioned Naval Officers and all studying for degrees in Engineering. It would be five years before I returned to the Fleet as a graduate Electrical Engineer.

****

For Christmas 1964, I opted out of a boozy College Bachelors' Club excursion and booked into a ski-ing course at Glenmore Lodge near Aviemore, an alcohol-free Outward Bound school run by the Scottish Council for Physical Recreation. This was no holiday camp. Ski-ing in Scotland at that time was still evolving from mountaineering and I would be spending most of the week carrying my skis uphill only to gain very short downhill runs.

On the train north, in Perth station to be exact, I met a slim, vivacious girl with an infectiously cheerful disposition and a mop of black, curly hair. She too was en route to Glenmore Lodge for the same ski-ing course. Her name was Catriona Thomson, almost one of the clan except that she spelt Thomson without a P. She came from Uddingston, only five miles from home, and was a newly qualified teacher. As we spoke the same language, I revived my dormant skill in Glaswegian jokes and she laughed. In fact, she laughed at them all. (Ladies, the way to a man's heart is not through his stomach; it is by laughing at his jokes). I was smitten. It was as if I had just found the right key for the front door.

At the first-night ceilidh in the Lodge, I asked her to partner me in *Strip the Willow*, a particularly energetic Scottish dance, during which our eyes met. Hers positively sparkled. Pure fizz. 'This,' I thought, 'is the sort of girl I could marry.' Romantics would say it was love at first sight but I was more pragmatic. She said she was dating a Chartered Accountant in Glasgow and, I presumed, was about to be married.

As I was thinking of volunteering for submarines, I had arranged that Christmas to join *Olympus* at Faslane. She was a diesel submarine and I was to take passage in her back to Plymouth. On arrival, I wrote immediately to Catriona and tried to impress her with comments like: *I'm writing this beneath the waves.* Indeed I was. A submarine is like an iceberg in that alongside the jetty, most of it is beneath the surface. There was no reply but it had been worth a try.

Six weeks later, a letter with an Uddingston postmark appeared in my College pigeonhole. It was from Catriona. She wrote as she behaved, with humour and enthusiasm. It was simply wonderful to read. She said she had much enjoyed meeting me and hoped we could meet again. That was it. Love at first letter. When I asked her why she had waited six weeks before replying, she said that she had not wanted me to think she was 'putting me on a pedestal.'

Gentlemen, never underestimate the guile of a woman. Catriona – Kate as she liked to be known – was the most honest, unselfish, non-manipulative woman a man could ever wish to meet, but even she could inflict misery and disappointment on me for six whole weeks just to play hard-to-get. At College, I kept my budding postal romance secret – I was still a member of the Bachelor's Club and had credibility to maintain.

The following Christmas, Kate came down to the College Ball. By then my parents had very generously funded a second-hand Riley sports saloon for my twenty-first birthday, a reasonably impressive vehicle for a young bachelor. In this, we drove the four hundred and eighty-four miles back to Glasgow and as we came off the M74 on the southern outskirts of the city, I pulled into the entrance to Daldowie Crematorium, the first available lay-by. There I asked her if she would like a P.

> Catriona Jane Thomson, spelt with no P,
> Fell deeply in love with a man of the sea.
> Her desire was not for a man such as he,
> But because in his name he included a P.

I had then to seek her father's permission for betrothal.

Calum, her father, a six-foot-three-inch, mild-mannered, chain-smoking, stick insect of a man was a crofter's son from the island of Lewis in the Outer Hebrides. He was a first language Gaelic speaker, had learned to speak English at school, and was headmaster of Rogerfield Primary in the notorious Easterhouse housing estate on the outskirts of Glasgow. He was also prodigiously well read. With great trepidation, I decided to beard him in the headmaster's office.

'I've come to ask for permission to marry your daughter,' I said nervously.

'I've been expecting this,' he replied. 'Of course you can. The girl is potty about you. She's been checking the mail every morning to see if there was a letter from you.'

Her mother took a different view. She was the daughter of a Glasgow lawyer and, like Calum, had taught in the slum areas of the city. 'I'm not having my daughter marrying a man who swears,' she said.

'Mrs Thomson,' I replied. 'I have never sworn in front of you.'

'Oh yes you have,' she snapped back. 'You said 'bloody'.'

Till then, I had never thought of bloody as a swear word. 'Bloody good hand' and 'bloody well done' were everyday phrases in the Navy.

Six months later, against both her mother's and the Navy's wishes, and to the disbelief of the Bachelors' Club, we were married. Since our first exchange of letters, there had never been any doubt about that.

For twenty-one years, I had believed that one day I would meet the right woman. I had and it was worth the wait. We were the perfect match. Many years later, one of my lifelong bachelor chums told me that he regarded Kate and me as the ideal married couple. It was one of greatest compliments I've ever received.

Marriage was my second re-birth. I had just made myself utterly responsible for another human being – for life. Everything I did and thought now had to be done for two, the most immediate thing being to pass my final exams. If I did not, I would be out of a job. For her part, Kate became my inspiration. She would carry me through all the trials and tribulations that lay ahead. Living ashore with her also removed the temptations of the College bars and allowed me to study at night without interruption. As a result, I graduated with Honours.

It is amazing how success breeds success. One month after graduating, Kate announced that she was pregnant. I had now qualified as an Electrical Officer, been promoted to Lieutenant, found a wife, acquired an Alsatian puppy, and was about to become a father. It was time to start submarine training.

CHAPTER 6

Welcome to Submarines

'You don't get a second chance to create a first impression.'

OSCAR WILDE

On Friday 20th December 1968, I drove from the Submarine School in Gosport to Faslane to join *Otter*, my first submarine. I was now entitled to a brand-new married quarter in Rhu, fully furnished and with fitted carpets – but not for long. Kate and I had hardly unpacked when a Civil Servant arrived with a tape measure and carpet scissors. 'Lieutenants are not entitled to fitted carpets,' he announced and proceeded to cut strips off the new carpets, leaving a six-inch gap between them and the walls. I had literally been cut down to size.

There is an old saying in the Navy: *Them that's keen get fell in previous* and I was so keen that I decided to pay a social call on *Otter* a few days early, just to let the team know I had arrived. That, I thought, would be a clear statement of intent. As I descended the short vertical ladder from the main access hatch, my nostrils savoured the characteristic submarine aroma of diesel oil, stale cooking and body odour.

Otter was modern. First commissioned six years previously, it weighed 2,000 tons, was 295 feet (90 m) long and had a diameter of 29 feet (8 m). Internally, *Otter* resembled a corridor train, having only one main deck and a single narrow passageway running from for'ard to aft, with a crew consisting of six officers and sixty-four ratings. The Wardroom was tiny, no more than a kitchen table surrounded on three sides by double or triple-decker bunks that also served as seats, the implication being that bunks could not be used when seats were required. The fourth side contained a small communal wardrobe, a cocktail cabinet and the ship's office – a kneehole desk complete with portable typewriter and one-drawer filing cabinet. This was to be my new home.

In submarines, messes have curtains, not doors. When I reached the Wardroom curtain, my heart was pounding. After seven years of training, the time for my big entry had finally arrived.

In fairness to the residents of *Otter's* Wardroom, they had not expected an unknown officer to whip back their curtain at lunchtime during a Christmas leave period. 'Who are you?' a voice demanded.

'I'm Lieutenant Thompson, the new Electrical Officer.'

'He's the new Electrical Officer!' a second voice echoed, then squealed with laughter.

'You're not due till the 3rd of January!' an Australian voice squawked. It was the officer I was due to relieve.

'You're not due until the 3rd of January!' the second voice repeated, laughing like a jackass.

'Go away. Come back on the 3rd of January.'

The curtain was then pulled shut in my face. How humiliating. I could hear them laughing. It was as if I had just stuck my head into a den of hyenas. Nothing had prepared me for that.

There is an old saying: *You don't get a second chance to create a first impression.* Heaven help me. I was about to be serving with this team for the next two years and I had already made a fool of myself. Should I whip the curtain back once more and demand entry? With the discretion of a wimp, I called cheerily, 'See you all on the third,' and hastily retreated, tail between my legs. Conformity has never been my strong point.

I had first visited the Gareloch as a small boy when Father drove me down to see the Reserve Fleet, which was moored there. It was paradise for a boy who loved warships. There were lines of battleships, aircraft carriers and cruisers, all with famous names, all veterans of the war, all now scrapped. More recently, I had visited again for a taster of submarine life in *Olympus,* berthed alongside the depot ship *Maidstone* at Faslane. Since then, Faslane had changed out of all recognition.

A large shore establishment had replaced the depot ship. Barrack blocks, workshops, warehouses, sports facilities, a floating dock, and the Polaris school had all been thrown up and surrounded by a maximum-security fence. New schools and married quarters estates had been built in nearby Helensburgh to house the large influx of naval families whilst over on Loch Long, a top secret armament depot had been carved out of the hillside for processing American Polaris missiles and British nuclear warheads.

The Submarine Service had now inherited responsibility for the country's Strategic Nuclear Deterrent and our new Polaris submarines, SSBNs in the jargon, had been nicknamed 'bombers'. *Resolution* had already returned from her first operational patrol, *Renown* having relieved her on station in our first nuclear baton change. *Repulse* was on her way up from Vickers Shipbuilders in Barrow-in-Furness and *Revenge* was being built at Cammell Laird's in Birkenhead. *Dreadnought, Valiant* and *Warspite,* our first three conventionally armed nuclear-powered hunter-killer submarines, known as SSNs, were also now fully operational and based at Faslane. Their role in wartime was to hunt and kill Soviet submarines.

Since Nagasaki, the world's nuclear arsenals had been building up. There were now an estimated twenty-two thousand nuclear warheads on the planet; the term 'overkill' was even creeping into strategic calculations. Though content to escalate their own nuclear holdings, the five nuclear powers (USA, USSR, Britain, France and China) actually shared a fear of uncontrolled nuclear proliferation and had agreed a Nuclear Non-Proliferation Treaty which sought to prevent the spread of nuclear weapons, encourage nuclear disarmament and promote the peaceful use of nuclear power amongst the one hundred and ninety Treaty signatories. Nuclear war, it seemed, was to be the prerogative of the big boys, but India, Pakistan, Israel and North Korea did not sign. A worldwide Nuclear Test Ban Treaty had also been agreed but had no impact on bomb manufacturing.

In the midst of all this nuclear activity, *Otter* seemed like a very minor player.

My overwhelming feeling on joining was one of profound apprehension. Could I maintain the heroic traditions of the Submarine Service? Would my technical ability be sufficient to deal with the equipment problems I would face? I had never wanted to be an Engineer; I had wanted to be a Seaman and captain my own ship. Engineering was my duty, not my métier. I felt like an imposter.

The gist of my turnover from the Australian was: 'No worries, mate. Your Chiefs will run the department. Have another beer.' He was a loud-mouthed, rough diamond who had been promoted from the lower deck and I did not share his concept of managing a department. Once I had proved to myself that I had mastered the job, I would relax and join the party, but not before. Twenty-four hours later, he was gone and I had the weight.

I was now responsible for all things electrical, including weapon systems, propulsion, main batteries, and all sensor and communication systems including the electronic kit in her periscopic masts (Radar, Radio and Electronic Warfare) in addition to the two sexed-up optical periscopes. To maintain all this, I had a department of five artificers and five mechanics and was one of the three heads of department, the other two being the Marine Engineer and the First Lieutenant, the latter being the senior Seaman officer and Deputy Commanding Officer.

Almost before I knew it, we had sailed.

At sea everything changed and *Otter* switched to being serious. Sailing down the Clyde on the surface in January with two large diesel engines sucking ice-cold air down the conning tower turned the Control Room into an Arctic wind tunnel. Now I understood why we had been issued with white, roll-neck, submarine sweaters and long woollen underwear. For an 08.00 departure, we were also in red lighting as it was still dark outside. Then 'Diving Stations' was piped on full main broadcast followed by a flurry of further orders and responses over the intercom. The Control Room was now our nerve-centre with shadowy figures moving about against a backdrop of small red indicator lights. Kerry Barr, the First Lieutenant, stood in the centre and pulled all the strings.

'Clear the bridge. Come below and shut the upper lid. I have the ship,' the Captain called to the Officer of the Watch on the bridge.

'Upper lid dry. Permission to shut the lower lid?' the Officer of the Watch reported, having descended into the conning tower.

'Open One, Two, Three, Five and Seven Main Vents.' (Four and Six main ballast tanks were filled with diesel oil). There was a loud sigh like a giant whale emptying its lungs as the air keeping us afloat was expelled from the ballast tanks and down we went. Apart from a slight downward angle, it was no more exciting than descending in a lift. A stream of reports then came in from all compartments to confirm there were no leaks. We had dived.

The six officers now went into continuous one-in-three watch-keeping, with two hours in every six being spent on watch. It meant that three-and-a-half-hours was the longest stretch I could have for sleep but that was not always possible if the Wardroom was being used for something else. It was exhausting. I learned to sleep in the sitting position. I fell asleep standing up. I even carried a bottle of smelling salts for when I felt myself falling asleep on watch, a most heinous crime.

On the surface we went solo on the bridge and that made life easier as the watch bill stretched to one-in-five. The challenge then was the cold and wet. On the surface, a submarine sits very low in the water and the tiny open-air watch-keeping space in the fin is often drenched as the bow cleaves its way through large waves. Kerry Barr was responsible for the watch-keeping roster and simply told me that I was to keep bridge watches. How ironic! It was my failure to meet the eyesight standards for bridge watch-keeping that had forced me to become an Engineer.

I was extremely happy doing bridge watches. I loved them. I had done them in *Barrosa* and was perfectly competent but resented being ordered to do them. I should have been asked, but asking would not have entered Kerry's head. He was a bulldog-like authoritarian and I was a rookie, so it would have been foolish to argue. (When the thick West Lothian accent of one of our lookouts defied his comprehension, he gave the man a white stick and said: 'If you see anything, just tap me on the shoulder and point.')

From the outset, it was clear that I would have no support from my brother Engineer, Nick Cromarty, a poisonous, flaxen-haired Cambridge graduate. More than any other officer, he seemed determined to make me look stupid at every opportunity – not difficult at that stage. Perhaps he was afraid I might upstage him. Perhaps he was undermining me to enhance his own prestige. Perhaps he was just nasty by nature.

Rodney Bolfrack, the Sonar Officer, was a Short Service officer who saw himself as the life-and-soul of the party, a complete Hooray Henry and not my cup of tea but he was to become my watch-keeping partner. Perhaps I misjudged him. As he bought a handgun in Malta and smuggled it home, he may have been destined for the Secret Service – or the other underworld.

Ian Tallant, the Navigator, was a fellow Scot. He was brilliant at his job and very focussed on his duties. There was mutual respect or at least a quiet understanding, but he knew on which side his bread was buttered. When I was made to look stupid, he joined the laughing chorus. Just before I joined, his Jaguar had been found burned out and abandoned by the roadside. Not his fault as he had been at sea at the time but the fault of an officer from another boat who had borrowed it without permission, written it off and not confessed. Exactly the sort of trouble *Otter* seemed to attract.

The most important person on board was, of course, Lieutenant Commander Gavin Douglas-Grant, the Captain, a gentrified Scot whose father had been a senior officer in the Navy. Judging by his accent he had been educated in an English Public School but at least his hair was flaming red. Like Kerry, he had served in older submarines without Electrical Officers and had even carried my responsibilities himself as a First Lieutenant. That made it difficult for me to impress him since he seemed to know my job better than I.

From first introduction, I felt uncomfortable with Douglas-Grant. I could sense that he didn't like the cut of my jib. Alas, I never seem to give a good first impression – I can never hide my naturally worried look. Perhaps he had been in the Wardroom when I made my premature social call and had already decided that I was a wally. Either way, I sensed no interest from him in developing me as an officer, but why should he? I was supposed to be the finished article. I may have been straight out of the egg but *Otter* was not an incubator; she was an operational front line submarine. Without doubt, Gavin Douglas-Grant was a highly competent Commanding Officer but through my somewhat puritanical eyes, seemed to have a playboy mentality. I felt like Mr Bean to his James Bond.

In the Royal Navy, Commanding Officers traditionally remain aloof from their officers but in submarines this is not the case. The Captain's cabin in *Otter* was no more than a bunk and kneepad of a desk across the narrow passageway from the Wardroom, and the Wardroom was the only place he could hold meetings. He also ate with us. When I made the cardinal mistake of sitting on his seat, there were howls of derision. I hadn't known about the sanctity of the Captain's seat but not knowing is never an excuse in submarines.

A second cultural difference between submarines and surface ships is that the submarine captain mans the periscope during an attack. He is the star of the show, and not some general manager. He is more like a fighter pilot than captain of an aircraft carrier.

Another new rule I had to learn was: *The Captain is always right.* I didn't entirely buy that. Yes, when he's barking out orders in the middle of an attack, one obeys without question, but human infallibility is a dubious principle. What bothered me most about Gavin Douglas-Grant was that he seemed to condone Cromarty and Bolfrack's attempts to make me look stupid, and Kerry Barr's contempt for my status as a head of department. I felt like a misfit, the butt of all jokes. It was a new and unnerving experience.

I recalled the motto on my primary school bankbook: *Look after the pennies and the pounds will look after themselves* and my Cub motto: *Do your best.* I clung to the belief that if I did both, I would eventually win the Captain's trust. And if that failed, I had my mother's maxim to fall back on: *'Please yourself, son, and at least you're pleasing somebody.'* Mother had attitude.

****

On going to sea after a maintenance period, it was standard practice to carry out 'angles and dangles' (emergency drills) and system testing. This included a deep dive in the Arran trench for which we went to Diving Stations and I closed up in the Motor Room. Being conscientious, I also went down into the Lower Motor Room to inspect for leaks. If a pipe down there were to burst, it would spray on to the main switchboard and short out our propulsion system, a potentially serious problem. It was loss of propulsion during such a test dive that sank USS *Thresher*.

Thresher was the first nuclear-powered submarine to be lost. Having just completed a major dockyard overhaul, her captain was under pressure to get her back to sea. The nuclear reactor systems had been quality assured to perfection but the secondary systems had not and corners had been cut. At the time of the sinking, *Thresher* was engaged in sea trials and was still carrying dockyard workers on board – as were the British *K13* and *Thetis*, which also sank during post-construction trials. Post-build and post-refit sea trials are technically the most dangerous times for submarines short of enemy action, for that is when faulty workmanship is exposed.

Thresher's test dive was intended to prove that the sea-connected systems could withstand full diving pressure. They could not. A cooling water pipe ruptured and at that depth a burst pipe does not leak; the sea comes in like a hydraulic laser. The incoming torrent shorted out the electrical switchboards which scrammed the reactor and denied propulsion. Rapidly the submarine became tail heavy. At 1000 feet (300 m), blowing air into a submarine's ballast tanks has virtually no effect. The nuclear submarine depends on propulsive power to drive up to safety and *Thresher*, having lost all propulsion, sank like a stone plunging downward, tail first, into the deep ocean abyss, accelerating all the way. The crew would have known they were doomed and would have had several minutes of knowing their fate. One shudders to think of the chaos and terror on board during that final, vertical, death dive.

Somewhere on the way down, the hull would have ruptured. This would have been such a cataclysmic event that the diesel oil in the inboard tanks would have exploded, just as it does in the cylinder of a diesel engine. That would have killed everyone. By the time three-and-a-half thousand tons of submarine hit the ocean floor seven thousand feet below (2000 m), it would have been travelling at over a hundred miles an hour, the remains being buried in the deep ocean mud. One hundred and twenty-nine men lost their lives. The tragedy had been caused by lack of Quality Assurance in her sea-connected systems and a *kamikaze* method of pressure testing.

I had learned *Otter*'s pipework systems thoroughly but in the forward starboard corner of the Lower Motor Room, there was a jungle of obscure pipes called the Sub Pressure System. It belonged to the Engine Room and was used only to ensure that diesel oil in the external fuel tanks remained at lower pressure than the sea, thus ensuring that oil did not leak out and compromise our position. From one of these pipes, a gentle fountain of water was reaching up to our vital switchboard wiring. It was only a pinhole leak but if it blew, we would be in trouble.

'Leak in the Lower Motor Room,' I yelled up to the switchboard operators who repeated the message on full main broadcast.

Prompt action was now required but I didn't know which valve to shut in the spaghetti-like conglomeration of pipework and did not want to make matters worse by shutting the wrong valve. As my most immediate concern was to protect the vital electrics, I took off my battledress jacket and, like Sir Walter Raleigh, laid it on top of the jet of water. At that point, a mighty hand was placed on my shoulder and the voice of Plugger Rawle, the Chief Engine Room Artificer, boomed out: 'What the fuck's going on, Sir?'

I felt utterly stupid. Plugger Rawle was King of the Engine Room. In fact, he was King of all things mechanical on board. During the War, when submarines did not carry Engineer Officers, the Chief ERA was top dog and reported directly to the Captain. Plugger had just witnessed a college-boy dealing with a submarine leak by laying his jacket over it.

'Out of my way, Sir,' he called in a voice one dared not disobey. That was the courtesy. In fact, he heaved me out of the way and shut the correct valve. Emergency over. It was my action that had spotted a potentially serious problem but the incident did nothing to improve my self-esteem.

****

I had virtually no dived watch-keeping experience when I joined *Otter* but that was not what had concerned me; my anxiety had been over my competence as an Engineer. The Captain took the opposite view. He didn't seem to give a toss about my technical competence. His priority was that I should be a competent watch-keeper and therein lay the seeds of my downfall.

In the North Atlantic in winter, a diesel submarine spends most of its time in dim red lighting, even when deep, to ensure that the Captain always has his night vision ready for coming up to periscope depth. I was Officer of the Watch in such dim red lighting when a main hydraulic failure occurred. We had lost all power to the hydroplanes and rudder. It was a real emergency. We were out of control. The main hydraulic pumps had failed. I scanned the array of dim red dials and lights before me and noticed that a number of small indicator lights were extinguished.

That was the clue. Immediately I diagnosed a main electrical power failure. That was why the pumps had not cut in.

I took the right recovery action but the fact that it had happened at all worried me as the electrical supplies were my responsibility. Then I discovered that in collusion with the Captain and First Lieutenant, Cromarty had gone back to the Motor Room and switched off power to the pumps. I was being tested. The Captain and First Lieutenant had been lurking in the shadows to see what I would do. I could hardly object to having my competence tested but what stuck in my gullet was the conspiracy element; they had ganged up on me; we were not in a training period; I had not been accepted. And how dare Cromarty go interfering with my power supplies? I would never have gone into his engine room and shut off the fuel supply. If I had, Plugger Rawle would have had my guts for garters.

I now saw a new side to life in submarines. In the few days I had spent in *Olympus*, humour had been the salient characteristic. That, I had presumed, was the submariner's survival mechanism. I was wrong. In *Otter*, the survival mechanism was to crack the new boy, the principle being that if a man will crack under peer group pressure, he will crack under enemy attack – different ships, different cap tallies. I was determined not to crack but felt completely isolated. I was on my own with no prospect of moral support, nor did I want any. This was a battle I had to fight alone. The big boys all seemed to be against me and they held all the cards. Or was I becoming paranoid? I was certainly under stress. I could hardly think straight, I was permanently pre-occupied, my concentration kept drifting and that made it even more difficult to remain alert on watch. What little self-confidence I had was rapidly vanishing.

This for me was a virility test. It was about proving that I was worthy of being a submariner; that I could survive in a man's world. I had to succeed. But would I be driven to a nervous breakdown before I did? It had happened to others; they were sent to Netley, the Royal Navy's Psychiatric Hospital. I was determined not to let it happen to me. How could I ever face Kate if that happened? I felt cornered. I was fighting for my professional survival.

Never mind the enemy, life in a submarine is tough enough without involving them. I read in a science magazine that an applied psychology experiment had shown that the more rats crammed into a small space, the greater the signs of stress they exhibited. It was hardly surprising therefore that when seventy men are crammed into a small space like *Otter* for long periods, stress occurs. Apart from testosterone, ambition, egotism, authoritarianism, haemorrhoids, God complexes, sleep denial, lack of oxygen, lack of female company, suppressed homosexuality, alcohol and nicotine withdrawal symptoms, distrust of others for your own safety, obnoxious personal issues like smelly feet and damned difficult jobs, there was also latent fear. A submarine was continuously at risk from so many different forms of hazard: collision, fire, flooding, grounding, sinking, poisonous gas, weapon or battery explosion, electrocution, and radioactivity as well in nuclear boats. All of these hazards were present without going to war.

Since the Second World War, the Royal Navy had lost two diesel submarines, *Affray* and *Truculent,* in inshore British waters. The French had lost *Minerve* and the Israelis *Dakar*, both in the Mediterranean. The Americans had lost *Thresher* and *Scorpion*, both nuclear-powered, in the Atlantic. During the course of the Cold War, the Soviet Navy lost about eight. Virtually all of these submarines were lost with all hands and their hulls never recovered, the cause of their disasters remaining unknown. The Soviet submarine *Kursk* was an exception. The wreck was recovered and torpedo detonation confirmed as the cause of the disaster; its bows had been blown off. Those were the sinkings, but there were many more non-sinking accidents, some due to material failure, others to human error. It was hardly surprising therefore that submarine crews developed psychological survival mechanisms.

In time of war, the risks go off the clock. During the Second World War, the Royal Navy lost the entire strength of its Submarine Flotilla at the outbreak of hostilities. Our submarines had operated exclusively in enemy waters in both World Wars and had also been the first British forces into action as they were already in their patrol positions. Thank God I was serving under a nuclear umbrella. I was hoping to keep war out of the equation.

****

In February, we were based in Londonderry, our role being to act as a target for Coastal Command's Anti-Submarine *Shackletons* based at RAF Ballykelly. We were to patrol the sea area between Donegal and the Outer Hebrides and every day the '*Shacklebats*' would lumber out to search for us and then go home for tea while we remained in full Atlantic conditions, waiting for the next day's attacks.

The troubles that plagued Northern Ireland in the late twentieth century had not yet begun and Londonderry was still a fun place to be. Our base was HMS *Stalker*, an old Tank Landing Craft, tied up conveniently alongside a main waterside street. That allowed the morning muster of the Duty Watch to be held across the road in Cassidy's Bar! As girls from the local shirt factories seemed ever ready to befriend a sailor, the crew were very happy. The officers fraternised with Wrens from HMS *Sea Eagle* or drank in Bridget's Maiden City Bar in the centre of the old town. The IRA destroyed it during The Troubles – hardly surprising as its walls were covered with the crests of visiting British warships. We also thought nothing of popping over the Border into the Irish Republic for a night out in the Squealing Pig at Muff. On our first day in, Kerry Barr issued me with a sheriff's badge and told me I was Duty Officer. That set the tone.

For the parties we held on board, I fitted a twelve-inch bass loudspeaker in the kneehole of the desk in the ship's office, one of my few acknowledged triumphs.

There was no need to fit red lighting as we already had that. A regular attender at our parties was a boisterous young Englishwoman called Becky Nickersoff who sounded like a talking Navy List. I had never heard such name-dropping. She seemed to know every officer in the Navy except me. At one party, she even filled her shoe with champagne and passed it round as a communal cup. I feared we would all go down with foot-and-mouth disease.

In this jocular environment, I felt ready for my first attempt at humour and slipped a plastic fried egg into Rodney Bolfrack's breakfast. In red lighting, egg, bacon and baked beans, are completely colourless and much to everyone's amusement, he attacked the plastic egg with his knife and fork. No one suspected me; I was far too serious for that.

On our way back to Faslane, we went up to Inverary at the head of Loch Fyne where we sat on the seabed at four hundred feet while staff from the Escape Training Tank conducted deep escape trials using our single man escape towers, tubular vertical pressure chambers inside which a man in an immersion suit is pressurised rapidly to sea pressure and then released like a human torpedo to shoot up to the surface. (All submariners practise this drill in the tank at HMS *Dolphin,* the Submarine Training School, but only down to one hundred feet). These brave men were pushing the boundaries at considerable risk to themselves and all reached the surface safely, a world record at the time. It meant we could escape from a sunken submarine from virtually anywhere on the continental shelf. Beyond that, in the deep ocean abyss, there would be no chance of escape. The hull would be crushed long before it reached the bottom.

On return to Faslane, I was responsible for managing our maintenance period and was particularly good at that. I had an excellent relationship with our crew and Base Staff and had the boat ready to sail as planned on April 1st. To mark the occasion, I smuggled a large leek on board and on our way down the Clyde, waited until the Captain had gone into the Wardroom for morning coffee. I then laid the leek on his bunk before going to the ship's main broadcast, switching out everywhere except the Wardroom, and piping: 'Leek in the Captain's cabin. Leek in the Captain's cabin.' (In submarines, orders are always repeated). Immediately, I restored the broadcast system to normal, went over to the chart table and looked busy.

The words 'leak' and 'flood' in a submarine are sacrosanct. 'Leak' means water is coming in; 'flood' means the boat is in danger of sinking. But we were safe on the surface and I had planned for that. Out of the corner of my eye, I saw the panicked rush from Wardroom to Captain's cabin, led by the Captain whose seat was next to the door. He was not amused. I doubt if the incident enhanced my career prospects but it did pass into submarine folklore.

****

In early May, we set off on a long deployment to the Mediterranean without Petty Officer Rusby, my radio maintainer. He had suffered a nervous breakdown. The surface passage to Gibraltar was in cruising routine with the weather improving all the way. We even changed into whites, or as white as our uniforms could be in a diesel submarine. Two things awaited us on arrival – a mailbag and the name-dropping Becky Nickersoff. The woman was either a Soviet spy or having an affair with one of us. The news from Kate was that she was pregnant again. I was thrilled about that but more worried than ever about my ability to support an enlarged family if my career failed.

When we sailed from Gibraltar, it was without my Leading Radio Mechanic, Rusby's assistant. He had missed the boat on sailing and was in serious disciplinary trouble. Now I had no radio staff and would have to handle any radio problems myself.

From Gibraltar, we headed for Hammamet in Tunisia to do some spying. The Soviet Navy was building up its Mediterranean Fleet and using Hammamet as an anchorage. This we were to investigate. As we approached, we sighted *Ugra*, a Soviet depot ship, on the move. Immediately, the Captain decided to conduct an under-run. That meant creeping slowly beneath and examining the bottom with a depth separation of only a few feet. How exciting! The hull was featureless with a rudder and propeller at one end, hardly the intelligence breakthrough of the century but it had everyone's adrenaline flowing and demonstrated the Captain's skill as a ship-handler. It was a seriously hazardous operation. If the Soviets detected us they would react aggressively and any slight miscalculation by the Captain could result in an underwater collision as had recently happened with *Warspite* during just such an under-run on a Soviet submarine.

Later, we bumped into *Ugra* again, metaphorically speaking. This time we were both on the surface and they tried to ram us. It was the sort of thing the Soviets did. The following year in the Mediterranean, a Soviet destroyer actually collided with the aircraft carrier *Ark Royal*, which she had been shadowing. Both ships sustained damage and the Soviets lost two crew members. This was not a game.

On surface passage from Hammamet to Malta, I was on watch on the bridge when Nick Cromarty joined me, a unique event as he was exempt from bridge watch-keeping and would certainly not have come up for a social chat. He was carrying a small transistor radio and a length of wire. The wire he proceeded to wind round our emergency whip aerial, a fishing rod type of assembly that we rigged on the surface.

'What on earth are you doing?' I asked.

'You wouldn't understand,' he said. 'We haven't cleared our surfacing signal. There's a problem with our aerials.' (If we failed to clear our surfacing signal, it would trigger a SUBMISS (submarine missing) emergency).

How dare he? This was my department. I called immediately for a relief and went below to investigate.

The Radio Supervisor informed me that our main periscopic HF radio mast and emergency whip aerial were both defective. He had reported this to the Captain but had not bothered to tell me. Normally, my radio maintainers would have been called and would have told me. But why had the Captain not sent for me? That was why I was there. Why had he gone to Cromarty?

This was the scenario I had always feared. With my radio specialists missing, it was now down to me to solve the problem. Having poured over electrical diagrams and measured insulations, I came to the conclusion that there was an earth fault on the main radio mast, the most likely explanation being that the mast was flooded. This would happen when a boat went too fast with it raised. Someone had broken the speed limit.

'I'll need to take the bottom off the mast,' I said to Cromarty in front of the Captain, Cromarty being responsible for the ironmongery of the mast.

'Don't be ridiculous,' he replied. 'You can't take a great heavy casting like that off at sea.' He had made me look stupid again.

Fortunately, our surfacing signal was cleared through VHF via a passing NATO aircraft. The panic was over but we still had defective HF aerials. On arrival in Malta, the Captain sent for me. It was the first time we had actually had a one-to-one. 'You are to remain on board until your aerials are working,' he said.

What? Being 'required on board' is a punishment in officer terms. I was livid. I was being punished for the radio masts failing!

My first action was to phone the RN Radio Station at Lascaris and ask for help. Within minutes, their team of experts confirmed my diagnosis and, as Cromarty had gone ashore, I removed the supposedly heavy casting from the bottom of the mast. It was not heavy at all and very easy to remove. Cromarty had been bullshitting. As the bottom came off, a column of seawater came tumbling out. The mast had indeed been flooded. My diagnosis had been perfectly correct but the Captain had preferred Cromarty's bullshit to my professional advice. I raged at the injustice.

Our programme was now delayed while a new mast was flown out from Faslane and, as Wardroom wine caterer, I made the most of the delay by doing a stock muster in the wine store – a tiny storage space entered via a manhole in the main passageway. During the muster, I realised that I had lost all power of concentration. I was counting bottles, could not find my stock book, could not find my pen, could not remember the number I had just counted and, after a recount, could not remember which brand it was. I was losing the plot. I was becoming dysfunctional. I had reached the end of my tether. Was this a nervous breakdown? That was the moment this worm turned. I had to take matters into my own hands. I chose the nuclear option. I wrote a formal letter to the Captain requesting to be relieved of my duties under his Command.

When the moment seemed right, I went to his cabin and handed him the letter. There was so little space that we were almost touching. He was sitting. I was standing. The curtain behind me was drawn. It was the definitive eyeball-to-eyeball interview, the most dramatic moment of my life so far. He read the letter, looked at me, and said: 'Relax, you're taking things far too seriously.' With that, he tore up my letter and threw it into his waste paper basket. That I had not expected. Perhaps things were not so bad after all.

****

On leaving Malta, we acted as target for some long-range sonar being developed by NATO. During the trial, I was on periscope watch whilst snorting (using a massive snorkel mast to supply air to the engines). A trawler had just passed down the starboard side but a few minutes later re-emerged through our cloud of exhaust fumes. It had turned and was pursuing us. Immediately I called the Captain.

In seconds, he had taken the periscope and ordered: 'Stop Snorting – Emergency Stations – Full Ahead Together – Flood Q – Down All Masts – Keep One Hundred and Fifty Feet.' Bedlam ensued. Engines were crash stopped, masts lowered, the Quick Flooding Tank (Q) in the bow was flooded and with planes to full dive, the boat took on a steep downward angle towards safe depth. As the periscope came down, the Captain glared at me and said: 'Buck up, Electrical Officer.' I should have initiated the emergency drill myself.

In the midst of this drama, Leading Mechanic Billy Smart wandered through the Control Room in a full polka dot clown outfit. He had unearthed it in a bale of industrial rags. Such was life in a diesel submarine.

After the sonar trial, we headed to Port Vendres in the South of France, a favourite spot for Charles Rennie Mackintosh, the famous Scottish artist/architect, and just round the corner from Collioure, one of Picasso's favourite haunts.

We were scarcely alongside when a fire was reported in the for'ard Seamen's Mess. I rushed for'ard and found white smoke seeping out of the large, grey, battery-breaker box in the corner. I smelt the unmistakeable odour of burning electrical insulation. When I opened the breaker box, an electrical arc was dancing merrily between the main battery terminal and the steel frame of the box. It was a full earth fault and extremely serious. One massive 440-volt submarine battery, with enough power to propel two thousand tons of submarine at fifteen knots, was trying to discharge its stored energy through this arc. It was my very own lightning strike. If not stopped, it would lead to a catastrophic fire and cripple the submarine – and 440-volts were lethal. Great care was required.

As water and electricity don't mix, I grabbed the nearest carbon dioxide fire extinguisher and blasted the arc with enough carbon dioxide to create a frozen shell over the live terminal. The arc was extinguished immediately but within seconds it was back. It had burned through the ice plug.

By this time, the Captain was welcoming French civic dignitaries on board, not a good time to initiate a full fire drill. Instead, I relied on my competence. The main battery was in two halves in parallel. If I disconnected the forward half, there would be no immediate effect on the rest of the boat. Then I could break a battery link to disrupt the flow of power to the arc. My plan worked and the Captain and his guests remained blissfully unaware of any problem. I felt very pleased with myself and my team were suitably impressed.

Later, in a local hostelry, I bumped into Plugger Rawle and the Second Donk Shop Horse who were enjoying the local vino. (In diesel submarines, the engines are called 'donks' and the artificers who look after them are called 'donk shop horses'. Our two donks even had brass tallies on their front ends proclaiming them to be *Castor* and *Pollux*, the twin brothers of Greek mythology who were the patrons of sailors. The legend being that they were born from an egg along with their twin sisters, Helen of Troy and Clytemnestra, but our twins had been born at Scott's shipyard in Greenock.)

It was the day of the Monaco Grand Prix and Graham Hill of Great Britain, the reigning world champion, had just won. When I remarked on this, Plugger informed me that Graham Hill had been a Naval apprentice and they had served together in the cruiser HMS *Sheffield*.

'Oh,' I said, 'then we must invite him on board for a drink.'

'And how would we be doing that?' Plugger asked, assuming I was joking.

Here was a chance to score credibility points with someone who really mattered. Plugger was a big man in every respect. During his career, he had served in every donk shop horse position in submarines and would also have been an Outside Wrecker, the man responsible for every piece of mechanical equipment outside the Engine Room. Plugger knew every nut and bolt on board. He was the guy who was carrying Cromarty. I had no such luxury. My department was a collection of singletons, each a specialist in a particular equipment. No one but no one tangled with Plugger, not even the Captain, and Cromarty would not have dared. If I could impress Plugger, I would be damn near anointed.

'I speak French,' I said. 'I'll get in touch with the British Embassy and find out where Graham is staying and invite him over.'

Plugger's jaw dropped.

I asked the barman if I could use his telephone and phoned the British Embassy in Paris. Graham Hill, they informed me, was staying at the Hotel de Paris in Nice. I phoned the hotel.

'S'il vous plait, je veux parler avec Monsieur Graham Hill,' I said in my schoolboy French.

'Mr Hill is at the Prize-winners' Dinner at the moment,' the receptionist answered in perfect English.

'I know,' I said, bluffing, 'but I have a very important message for him. Will you please tell him that I'm calling from a British submarine and have one of his friends from the Navy with me? He's invited on board for a drink.' The receptionist probably thought we were in Monte Carlo harbour.

I was put on hold for a few minutes then a voice said: 'Hello, Graham Hill speaking.'

Wow! I was actually speaking to the reigning World Champion. I signed to Plugger to come over immediately.

'Good evening, Mr Hill,' I gasped, 'congratulations on winning today. I have Plugger Rawle here for you.' With that I handed the phone to Plugger.

The scene that followed was pure joy to behold. Our mighty Chief ERA visibly shrank into a hero-worshipping schoolboy. He talked with Graham for about five minutes and when he came off the phone, looked utterly dazed.

'That was Graham Hill,' he said, as if in a dream. 'He remembered me.'

I had just shot to the top of Plugger's pops. Beat that Cromarty, I thought.

I left the pair to enjoy their vino and, boy, did they do that. So drunk did they become that the Second Donk Shop Horse lost his false teeth and remained toothless until we returned to Faslane.

It was simply wonderful to know that a man of Graham Hill's superstar status was prepared to leave the top table at his victory banquet to speak to some old acquaintance from his Navy days. There is something truly spiritual in the bonding between old shipmates. Tragically, Graham was killed in an air crash six years later.

During our passage back to Gibraltar, Ian Tallant came into the Wardroom and said casually: 'Bad news for somebody, chaps. There's an Officer's Report on the Captain's desk and it's all underlined in red.' It had to be mine.

We arrived back in Faslane five days later without further incident. On the eve of arrival, the Captain called me into his cabin and drew the curtain.

'Your relief will be joining when we get alongside,' he said.

The statement hit me like a bullet between the eyes. I had been sacked! Short of gross misconduct, I could sink no lower. Seven years in training and after only six months, I had been sacked. I was consumed with shock and shame. It was exactly what I had sought to avoid with my letter in Malta. For the past four weeks, I had been living in a fool's paradise. I felt betrayed.

Now I had to go home and tell Kate that her husband was a failure. Of course, she stood by me like the rock she was but I never did tell my parents. They had been so proud of me. I could not disillusion them.

My time in *Otter* had been an unmitigated disaster. My career was over. My self-esteem had been virtually destroyed, but I counted my blessings. This rude awakening was the best thing that could have happened to me and it would serve me well in the future. In those few seconds in the Captain's cabin, I grew up. In future if I were to fail, I resolved to go down on my own mistakes, not on the mistakes of others. And never again would I allow a formal letter of mine to be torn up and thrown into a waste paper basket without it going on record. Never again would I allow a superior to overrule my professional judgement without a written agreement to that effect.

One lesson I did not need to learn: I would never treat a subordinate of mine the way my Captain had treated me. He was brilliant at driving his submarine, a veritable war-hero-in-waiting, but had failed to inspire me.

As I trudged home to tell Kate, it seemed as if I were a million miles away from the world of nuclear-powered submarines and strategic nuclear deterrence. I felt as if I were the only failure in the entire Royal Navy. Later I would discover that I was not. Most people face a crisis at some point in their careers. I had in fact been very lucky; I was so junior in rank that Gavin Douglas-Grant's adverse reports would not be included in any consideration for future promotion.

Subsequently, I discovered that Nick Cromarty had had a nervous breakdown in his first submarine, *Otter* being his second. Making me look stupid was his perverted attempt to win approval. It is always the bully who has the problem. I never heard of him again.

Until *Otter*, I had known only success. I had been head boy at school, won a coveted scholarship to Dartmouth, enjoyed a highly successful year as a Midshipman, earned my Commission, and fought my way to an Honours degree in Electrical Engineering. I was in the crème de la crème of the officer corps, then crash, bang, wallop, I was a failure. Why had things gone so horribly wrong? The answer is lack of self-confidence.

My black bushy beard plus a red haired Captain also seemed to have been a catastrophic mix; so I shaved off the beard. I had never wanted it in the first place. I had simply tried to look the part.

Otter had been a truly character-forming experience and I left knowing that I had proved myself as an Engineer. On that at least, I could hold my head high.

CHAPTER 7

Resurrection

'If at first you don't succeed, try, try, again.'

PROVERB

I was now a reject. Douglas-Grant had washed his hands of me. The Squadron had turned its back. No one was interested. When I phoned my Appointer, he told me I would be going back to General Service.

'I don't want to go back to General Service,' I replied, shocked. 'I volunteered for submarines. If I can't do that, I wish to leave the Navy.'

'Oh,' he said. 'I thought you didn't want to remain in submarines.' Who told him that? I had never said that. No one had asked me what I wanted to do.

A few days later he phoned to say he was sending me to HMS *Andrew* in the Second Submarine Division based in Plymouth. Second Division! It felt like I'd been relegated to non-league football – and still had to prove I was good enough for that. Much as I loved Plymouth, this was a domestic disaster; we had just bought a house in Helensburgh, five hundred miles away, and Kate was three months pregnant. She would have to cope on her own but she would have done anything to help me.

Andrew had just returned from Australia and was smaller and much older than *Otter*. She was a relic of the Second World War. She even had a 4-inch gun on the casing – in the age of Intercontinental Ballistic Missiles. She had been designed for high-speed surface transits across the Indian Ocean from Trincomalee in Sri Lanka to the Japanese occupied waters around our Far Eastern colonies, her massive diesels driving the propellers directly. That made her fast on the surface but far too noisy to be of much value in the Cold War. Her batteries were also half the voltage of *Otter's*, which gave her much lower underwater speed and endurance. It felt as if I were travelling backwards into submarine history.

When I reported on board, the social differences from *Otter* were immediately apparent. Peter Emerson, the First Lieutenant, welcomed me most warmly. Although our paths had never crossed, he had joined the Navy with me, was a devout Christian and shortly afterwards would resign from the Navy to become a missionary in Africa.

His two subordinate Seaman Officers were on Short Service Commissions: Stuart George Ellson, the Sonar Officer, would go on to become a bigwig in the Ministry of Agriculture and Fisheries, and Alan Prince, the Navigator, became a West Country thatcher at the end of his five years. The Marine Engineer, Malcolm Butler, a hearty Welshman, had been serving in the Royal Yacht as an Artificer and had just been promoted to Sub Lieutenant. They knew nothing of my *Otter* experience but must have viewed me with mild curiosity. For a Manadon-educated, career Electrical Officer to be parachuted into an old *A-boat* that had survived the previous twenty-five years without one must have seemed strange. *Andrew*'s electrical switchgear was out of the Ark. It was an open array of burnished copper knife-switches, the big ones being pulled open manually using string lanyards for safety, which was not the sort of stuff that required a graduate engineer.

Enter Richard Tobias Frere, my new Commanding Officer. He had brought the boat back from Australia, was still a Lieutenant, and was on honeymoon when I joined. On return, he simply welcomed me as one of his officers. He would, no doubt, have been briefed on my troubles in *Otter* but made no mention of it and never made me feel under threat. He was to be my personal saviour and as history would relate, our paths would intertwine successfully hereafter. Toby was a natural team-builder, quietly spoken, even-tempered, and had an inscrutable smile. He sought consensus and had acute political antennae that would take him to the rank of Fourth Sea Lord and a knighthood.

****

Andrew's first deployment was to Wales. We were to be one of the many RN ships visiting Welsh ports for the investiture of the Prince of Wales at Caernarfon Castle (July 1969). As we had an element of choice in our destination, the First Lieutenant assembled us in the Wardroom. The Navigator laid a chart on the table and the Captain asked for our preferences. It was a new experience to be asked for my opinion, but that was how Toby Frere managed things.

Our choice was Abersoch in Cardigan Bay, not a good place for a submarine because the bay was shallow and we had to anchor a mile offshore, but it was brilliant for the crew who were billeted in Butlin's holiday camp at nearby Pwllheli. We were to be Guard Ship for the South Caernarvonshire Yacht Club's regatta, the Club boat ferrying us back and forth. On one such trip, we dropped off a particularly loud-mouthed yacht owner at his yacht. I had seen and heard him in the clubhouse. Clearly, he regarded himself as a bigwig or else he was deaf.

The following evening, whilst hosting a cocktail party on board, I was asked to go to the Wardroom as the Captain wished to see me. I found him talking to the loud-mouthed bigwig.

'Eric, who made our switchboard?' he asked.

'Whipp and Bourne of Rochdale,' I answered, quick as a flash. Thank goodness I had paid attention to detail in familiarising myself with *Andrew's* antiquated equipment.

'This,' said Toby, 'is Mr Whipp. Could you take him back aft and show him our switchboard?'

The switchboard was really an extension of the Engine Room and consisted of two banks of large, exposed, copper switches, one either side, with wooden rails to prevent personnel falling on to the live copper in rough weather. Between the banks of switches was a bench seat for the watch-keepers. Mr Whipp and I sat down beside each other, brandy glasses in hand.

'My brother designed all this,' he said quietly, wiping a tear from his eye. 'He was electrocuted while testing switchgear for your Polaris submarines.'

A very large penny dropped. Our array of copper switches meant far more to Mr Whipp than to any of us. Suddenly, I had a completely different take on this 'loud-mouthed bigwig.' He was utterly charming and his family firm had done much more to serve the Royal Navy than I. He was a fully paid up member of Team GB, albeit slightly hard of hearing, and his brother had given his life in servicing the needs of the Navy.

Almost a year later when *Andrew* was due to visit Manchester, I wrote to Mr Whipp to invite any of his older employees who might have had a hand in manufacturing our switchboard to come and view it after almost thirty years in service. I had expected at most a carload but the invitation proved so popular that the Company had to ballot for seats on two forty-seater coaches. We couldn't actually handle that many, but I didn't want to spoil their party.

In the event, I stopped all leave for my small department, divided the visitors into manageable packets and gave each a guided tour. Their appreciation was palpable. In return, Mr Whipp who had come with them, invited my entire department back to Rochdale for a slap-up banquet in a five star hotel. The lads loved it. Mr Whipp, in excellent form, presided. Everything was at his expense including as much booze as we could consume.

In the middle of the banquet, the double doors of the dining room burst open and a formidable lady came barging in. She reminded me of Lady Bracknell in Oscar Wilde's *The Importance of Being Earnest*. She stood in silence for a moment, hands on hips, radiating displeasure. 'Frank!' she called imperiously, looking daggers at Mr Whipp.

We held our breath.

'Frank,' she repeated, 'where's your teeth?'

It was Mrs Whipp.

The following morning, after a tour of the factory, I called in at Mr Whipp's office to thank him. There, I noticed the latest copy of *Jane's Fighting Ships* on his bookshelf, an encyclopaedia of all the world's warships. 'Ah,' I said, 'I chose the 1961 edition for my Leadership Prize at school.'

He was on his feet immediately. 'Then have the up-to-date one'. He pulled it from the shelf and presented it to me. It still sits on my bookshelf at home.

From Abersoch, we headed north to the Clyde for Perisher running. In this, trainee submarine Commanding Officers are tested on their periscope attacking skills against four frigates simultaneously. This was not simulation. The frigates knew where we were and it was their job to run us down at full speed if they caught sight of our periscope. The risk of a potentially fatal collision was real and no holds were barred. The trainees had to place themselves in the path of an attacking frigate, hold fire for as long as possible, launch their imaginary torpedoes from point-blank range (a few hundred yards) and then duck down under the onrushing frigate before we were rammed. It was a highly professional game of underwater chicken. We had two-thousand-tons of frigate bearing down on us at twenty-eight knots and the trainees were under intense pressure not to chicken-out. Teacher was testing their nerve.

During the attacks, Teacher took over responsibility as Commanding Officer and sat on the large Main Periscope to ensure our safety, the students conducting their attacks on the slim, monocular Attack Periscope. If Teacher considered the situation was becoming unsafe, he would intervene and order the submarine to go deep in emergency. If he judged any student to be unsafe, that would be the end of that officer's submarine career. There were no half measures. That was why the Course was called the Perisher.

From my own watch-keeping point of view, it was also make or break. The trainees could not afford to have their attacks ruined through some basic mistake by a ship's officer. That was when I finally gained confidence as a Control Room watch-keeper. It also gained me recognition amongst Commanding Officers of the future.

****

At the end of Perisher running, we popped into Campbeltown to give the crew a run ashore. It is an unusual place. It sits at the end of the remote Kintyre Peninsula, one hundred and forty miles south from Glasgow by twisting roads. It is virtually an island, far more so than the islands of the Clyde, which have regular ferry services. Yet Campbeltown had an industrial feel to it. It seemed more like a sawn-off piece of Glasgow than a fishing village. There were tenement buildings instead of the cottages usually associated with fishing communities. As well as fishing, there was a small shipbuilding yard, an international standard golf course and the Springbank distillery, one of the world's top whisky brands. In his farm nearby, Paul McCartney would write the best-selling hit, *Mull Of Kintyre*.

Because of its proximity to the Atlantic, Campbeltown had been an important base for the Navy during the war. It was currently home to a massive NATO airbase, NATO oil fuel depot and was far more cosmopolitan than rural towns of comparable size (5,000 inhabitants) in Middle England. This also meant disproportionate trouble.

It happened to be Glasgow Fair Friday when we arrived. All the Glasgow factories and shipyards had just shut down for their summer holidays and so a large gang of Glaswegian bikers arrived at about the same time as *Andrew*. I was Duty Officer that night and all except the Duty Watch went ashore. The officers were back first at about eleven o'clock and quickly turned-in.

Shortly afterwards, the trot sentry appeared at the Wardroom curtain. 'Come quick, Sir. The police want you.'

An odd request, I thought, but it seemed urgent. When I arrived on the casing there was a police van on the jetty and a Sergeant standing by our plank (gangway).

'We want that man,' called the Sergeant.

'Which man?'

'The one in the water.'

'I don't see anyone in the water.'

'He's under the pier,' the Sergeant shouted.

'OK. You can have him when we get him out.'

By this time, Petty Officer Craig, the Duty Petty Officer, was at my side. 'Don't worry, Sir. I'll get him back,' he said, removing his shoes and jumping into the water. He then disappeared under the pier. Now I could have two drownings on my hands.

'What happened?' I asked the trot sentry.

'It was LME Jenkins, Sir. He challenged a group of Glasgow bikers to a fight in the local dance hall. The police were called and chased him back to the boat. He hit the plank at such speed, it tipped over and he went into the 'oggin.'

As we had no searchlight, I called for the signalling lantern to be brought down from the bridge but its lead was too short. We could only illuminate a barnacle-encrusted horizontal spar that ran just below the surface at the outer edge of the pier.

By now the rest of the crew were returning, mostly drunk.

'What'z the matter, Zir?' Petty Officer Booth, the Radio Supervisor, asked.

'We've got two men in the water.' I said. 'They're under the pier.'

'No problem, Zir, I'm the Ship'z zwimmer.' He whipped off his shoes and dived overboard.

Not another one! Before I knew it, drunken sailors were jumping into the water en masse. There was an instant water carnival. The place was beginning to look like a public swimming pool on August bank holiday. Then the Chief ERA and Coxswain arrived, our two most senior non-commissioned officers. They'd both had a skin-full but their sense of responsibility seemed intact. 'Get that bloody lot out of the water and below decks,' I roared at them. 'I'll deal with the guys under the pier.'

I wondered if I should call the Captain but what good would that do? He was no more able to control a bunch of drunks than I. It would only put him in an embarrassing position and make me look weak. I had to handle this myself and take what blame was coming.

About this time, in the loom of the signal lantern, three half-drowned figures appeared under the pier. It was the two Petty Officers with Jenkins slung between them, the latter looking all in. They had reached the barnacle-encrusted spar and for some inexplicable reason, the two Petty Officers seemed to be fighting each other. I watched as they submerged and re-appeared on the submarine side of the spar. Now they could grab the ropes we had thrown down.

It is impossible to climb on to the ballast tanks of a submarine from the water. They are round and smooth with no handholds, but the *A-Class* at least had flattish tops. By now, the entire Duty Watch was on deck helping me. They went down on to the tank tops and managed to haul out the three men, as well as the other water babies. We brought Jenkins up first, dazed, waterlogged, and drunk, but as he was a Welsh boyo and always up for a fight, he made an instant recovery. While we were hauling the two Petty Officers back on board, Jenkins was back on his feet, over the plank, and doing a runner back into Campbeltown.

'We want that man,' the Sergeant called, jumping into his van and giving chase.

I left the police to deal with Jenkins and went below, mightily relieved that no one had been drowned. In the narrow passageway through the Accommodation Space, the Coxswain greeted me: 'You'd better come quick, Sir. There's a fight in the Control Room.'

'What?'

For the Coxswain to seek my help was a serious admission of defeat since he was responsible for discipline on board. Unknown to me, Petty Officer Craig, a quiet, conscientious Ulsterman, and Petty Officer Booth, a loud-mouthed, bullying sort, were not friends. Their fight in the water had been because Craig had been managing Jenkins perfectly well under the pier until the drunken Booth arrived and had tried to take over.

When I reached the Control Room, the two were circling each other between the periscopes, fists raised. Fighting on board HM Ships is a very serious disciplinary offence, even more so for Petty Officers. They could both be dis-rated and sent to RN Detention Quarters. It worried me that the Coxswain had not been able to intervene. That was his job.

The golden rule for officers is never to get within striking distance of a drunken sailor because striking an officer is a court martial offence that would lead to the man being dismissed from the Service, thus aggravating the simple offence of drunkenness. I had no other option but to step between them.

'Gentlemen,' I said, 'my congratulations to you. That was very brave. Now don't let's spoil a job well done with unnecessary trouble. You both deserve a whisky.' They lowered their fists, a bottle was brought from the Wardroom and I gave them each a tot. When they had gulped it down, I said: 'Now get turned-in and we'll forget all about this. We're sailing in a few hours.'

As I could do nothing about Jenkins, I turned-in as well but had hardly hit my bunk when the Coxswain was back at my side. 'Come quick, Sir. They're fighting again in the Senior Rates mess.'

How very disappointing, Senior Rates should have been able to sort out such matters without involving an officer. I went round to their mess, a room about the size of a small bathroom with triple-decker bunks on all sides. 'That's enough,' I shouted in my best parade ground voice. 'Coxswain, put Petty Officer Craig into the Sound Room with a sentry and keep Petty Officer Booth here under your observation.' To my great relief, this worked.

My approach typified the difference in discipline between submarines and the surface navy. Had such an incident occurred in an aircraft carrier, the ship's policemen would have been called, the offenders would have been charged, locked up in the on-board cells and full disciplinary action would have followed. In submarines, the instinct was just to sort it out.

I had barely returned to the Wardroom when the trot sentry re-appeared. 'Come quick, Sir, the police want to see you again.' They had failed to find Jenkins.

'We want that man and this submarine is not to sail until we have him,' the Sergeant bellowed.

This was a serious turn of events. I would now have to waken the Captain and explain that the police had just impounded his submarine. As Duty Officer, I might as well go hang myself.

'Give me half an hour,' I called to the Sergeant. 'I'll send out a search party.'

I was now in dangerous territory. The only sober sailors at my disposal were the duty watch, all twelve of them, and they were required on board for safety and security. I took a gamble. I assembled them and asked for eight volunteers to go ashore, find Jenkins, and be back within half-an-hour.

Twenty minutes later, the trot sentry was back at my bunk. 'The police want to see you again.'

This time the Sergeant ventured on to the casing. 'Right, Sir,' he said, 'first, we've got your man. He's under arrest. Second, while we were arresting him and the Glasgow boys were all jeering, your other lads arrived and threw four of their motorbikes into the harbour. We've arrested them all and they're also in police custody. The submarine is not to sail until we have signed confessions.'

My God! The submarine impounded with fifteen per cent of its crew in clink. I had lost two-thirds of my duty watch and we were due to sail in six hours. The Captain had to be told. This time my career really was over. Heart in mouth, I went below, clambered up the Conning Tower ladder to his tiny cabin in the fin, a private pressure-tight chamber peculiar to *A-boats*, and gave him a shake.

'I'm sorry, Sir, but nine of the crew have been arrested and the police won't let us sail until they've sorted out the charges.' I stood ready for a rocket. What sort of Duty Officer was I?

Typically, Toby Frere was unfazed. He got out of his bunk, came with me to the casing, greeted the Sergeant most cordially and accompanied him to the police station. With his characteristic diplomacy, he got the required confessions. The culprits were charged, the nine men were released and we sailed on time.

The case, however, rumbled on for the next six months. The police had simply left the salt-water saturated motorbikes in their compound to rust beyond economic repair and the bikers' insurance companies were suing our bike-throwing culprits for the cost of four brand-new motorbikes. As luck had it, I was Divisional Officer for two of them and they appealed to me to defend them. I did. My argument was that had the police flushed the bikes with fresh water immediately after recovery from the sea, the bikes would not have corroded. It was therefore, I argued, police negligence that had caused the bikes to become write-offs and not the act of throwing them into the sea. My clients, I also pointed out, had been prevented from recovering the bikes as the police had arrested them and locked them up. Once released, the boat had sailed. I won the case. I should have been a lawyer.

****

From Campbeltown, we made the short hop down the Irish Sea to Belfast. How things had changed in Northern Ireland in the six months since my visit to Londonderry in *Otter*. There had been serious rioting following a recent Apprentice Boys of Londonderry march and British troops had been deployed on the streets to maintain law and order. The Troubles had started. For the next thirty years, a bloody terrorist war would rage in the Province with atrocities also being carried out in mainland Britain and elsewhere. Within a year, twelve thousand British soldiers would be deployed to Northern Ireland, initially to keep the peace but only to be branded as 'legitimate targets' by the IRA and its various competing factions. British servicemen would no longer be allowed to walk the streets of Britain in uniform lest they be assassinated. *Andrew* must have been one of the last HM Ships to visit Northern Ireland for a recreational visit.

On passage back to Plymouth, we loitered off the Lizard peninsula in Cornwall to act as target for new Sea King Anti-Submarine Helicopters based in the Royal Naval Air Station at Culdrose. These beasts were massive and designed to deliver the killer blow against fast-moving Soviet submarines. It was the first time they had exercised with a real submarine and to mark the occasion, the first helicopter lowered a box of Sea King ties as a memento. But that was not the only first of the day.

I was on low-stress periscope watch, quietly surveying the Cornish coast, when news came in that man had landed on the Moon. *Apollo 11* had made it. Better still, the first human to set foot on the moon was a Naval Officer – Neil Armstrong was a US Navy pilot. There was I sunbathing through the periscope in an antique submarine while up there on the Moon was a blood brother. Yes, I was definitely playing non-league football. That night, I surveyed the Moon in great detail through our high-powered Search Periscope and tried to imagine another human being up there on its surface.

****

Andrew was a maid-of-all-work. Our next adventure involved towing a large, hydrogen-filled barrage balloon with a radar reflector attached. This enabled scientists from the Admiralty Underwater Weapons Establishment to track us with an experimental, long-range sonar fitted in the frigate HMS *Verulam*. Our range whilst underwater was measured by radar reflections from the balloon. The balloon was attached to an Oropesa float, a torpedo-like body that 'flew' horizontally in the water like an aquatic kite. It was used in minesweeping for spreading towed cables. This was bizarre. Imagine a submarine on the surface being pursued by a torpedo with an airship hovering overhead. Then imagine the submarine dived. The viewer would see only an airship being towed by a small torpedo.

To the west of the Scillies, we encountered a fleet of Polish fishing boats. I was on periscope watch and saw one of the Poles detach from the fleet and come over to investigate. It was not a fishing boat. It was a Soviet spy trawler that had been hiding amongst the fishing boats. Goodness knows what it reported back to Moscow: *The British have invented a new secret weapon, a torpedo that tows an airship!* That must have reverberated through the corridors of Soviet Naval Intelligence.

Balloon disposal at the end of the trial seemed to have slipped under the scientists' metaphorical radar. They had no plan for recovering it. As our ultimate destination was Faslane, we now had to tow our giant balloon up the Irish Sea and into the Clyde Estuary. Toby Frere, a qualified pilot, dutifully sent a NOTAM signal (Notice to Airmen) to the Ops Room at Faslane for onward transmission to the aviation world – a large hydrogen-filled balloon was a hazard to aircraft. The Ops Room at Faslane, however, dealt only in submarines and ignored the signal. Thus, a V-bomber designed to fly our atomic bombs into Russia, whilst conducting a low-level bombing run up the Irish Sea, came face-to-face with a large unmanned airship. That almost started a Third World War between the Royal Air Force and the Royal Navy.

We were then refused permission to bring our balloon into Faslane on explosive safety grounds but could not simply release it as it would have risen to high altitude, been a further hazard to aircraft and its radar reflector could have been picked up on the Fylingdales Early Warning System and be classed either as an Unidentified Flying Object or an incoming Soviet ballistic missile.

Just off Arran, while I was on watch, a General Purpose Machine Gun and a clip of tracer bullets were sent to the bridge with instructions to shoot down the balloon. One shot later, I had a re-run of the *Hindenburg* disaster unfolding five hundred feet above my head. The balloon simply burst into a massive hydrogen-fuelled fireball, disintegrated, and fell burning into the sea. It was a most magnificent spectacle but witnessed only by a flight of passing gannets, the lookout, and me. A shame, for it was a true Hollywood spectacle. But how sad to see our beautiful balloon coming to such a sticky end; it had been with us for weeks and was almost like a pet. It had certainly put us one up on any other submarine.

On our way into Faslane, we passed the entrance to the Holy Loch where the Americans were operating their squadron of Polaris submarines, supported by the depot ship *Hunley*, named after the Confederate submarine *Hunley* that fought in the American Civil War. The name is significant in submarine history. *Hunley* was the first submarine to sink an enemy ship. In 1864, she attacked and sank the Federal sloop *Housatonic* but sank herself in the process, killing her crew, which was hardly surprising considering torpedoes had not yet been invented and *Hunley*'s method of attack had been to mount an explosive charge on the bowsprit and ram it into the enemy's hull. It was like sticking a bomb on the tusk of a narwhal and training it to head-butt its target. So the first submarine attack in history was actually *kamikaze*. I mused over how far submarines had advanced in a mere one hundred and five years.

When we returned to Plymouth, Toby Frere left us. I had served him for only three months but under his leadership, I saw a glimmer of light at the end of the tunnel. He had never criticised me. It was far too soon to assume an official 'all clear' and I was counting no chickens, but at least I felt that I was not entirely unacceptable in submarines. My future would depend on his successor.

****

Lieutenant Commander Michael Tuohy, Royal Navy, our new Commanding Officer, turned out to be one of the greatest characters (or lovable rebels) in the Navy. He was a tall, emaciated Anglo-Irishman with a highly cultured manner of speaking, a serious intellect, an unquenchable thirst for whisky, and a permanent readiness for mischief, especially if it involved putting one over on the Establishment or another boat. Immediately, he nicknamed me Elsan, a fictional Japanese Electrical Officer with the *double-entendre* of being a well-known brand of chemical toilet.

It took us no time to realise that in social terms, the Wardroom had just been turned on its head. From now on, our concern would no longer be whether or not we had pleased the boss; our concern now was in what manner of mischief our Captain was currently engaged. At sea, he was a master of the submarine attack and duly went on to become Teacher for trainee Commanding Officers. Ashore, he needed a minder. We had to organise a secret duty roster of officers to keep tabs on him. He was the smoothest reprobate in Christendom.

His reputation ran before him. A fellow officer had once met him coming out of the Gibraltar Casino. 'I guess that's the Beer Fund gone,' the officer had said in jest.

Tuohy smiled back. 'And the Welfare Fund.'

One never knew if he was joking.

Not long after he joined, his wife phoned the boat at about seven o'clock in the evening. I took the call. 'Has Michael left yet?' she asked. I think she was about to put his dinner in the oven.

'Yes,' I replied. 'He left to go into the dockyard.' He had actually left at midday and I had no idea where he had gone but it was an honest answer. Going into the dockyard was an essential pre-requisite to accessing the rest of the world. Later, I discovered he had gone on a pub and betting shop crawl. Being a good Irishman, he enjoyed both a glass and a flutter.

When relating this tale to a Squadron Officer, he looked pained. 'I share an office,' he said, 'with someone who tells his wife he's off to sea for the week in one of the boats. He arrives on Monday morning with his grip packed but is actually off to spend the week with his mistress. When his wife phones at the end of the week to ask when the boat's due back, he expects me to lie for him.'

When Mike Tuohy raised the periscope on his first dive, his lanky frame reminded me of a giraffe bending down to drink. He was visibly trembling with nerves. Submarine Commanding Officers are so well honed for their task it is easy to forget the weight of responsibility on their shoulders. Submarines go up and down as well as left and right and they fail-dangerous, the opposite of fail-safe. There are also between sixty and one hundred-and-fifty trusting souls on board and if it's a missile submarine, a nuclear reactor and boatload of intercontinental ballistic missiles with nuclear warheads to consider as well. Before even beginning to think about the enemy, there is much to keep a submarine Commanding Officer awake at night. That November, the *Thresher* Class submarine USS *Gato*, on an intelligence gathering patrol in the Barents Sea, collided with the Soviet missile submarine *K19* of 'Widowmaker' fame, at a depth of two hundred feet. Submarine collisions were a feature of the Cold War. Commanding Officers required cool heads. Mike Tuohy may have been full of mischief but there was no nonsense when it came to his Command responsibilities.

It was about this time that one of our outstanding young Commanding Officers ordered his submarine to surface and signalled that he no longer felt safe in Command and wished to be relieved of his responsibilities. That was a most morally courageous and hugely responsible action but I was shocked. I knew him. He was a most impressive officer and, I had presumed, was on his way to the top, but he had just committed career suicide. With a sense of responsibility like that, he was the sort of man who could be trusted. It is the arrogant, devil-may-care types who make me nervous. Fortunately, such types would not pass Perisher nor be selected as submarine Commanding Officers, not in the Nuclear Age.

The next few months were spent clockwork mouse running in the Portland Exercise Areas, our main concern being to avoid collision when at periscope depth. We were operating virtually in the middle of the English Channel with an array of warships from different NATO countries practising attacks against us but The Channel was also like a motorway for merchant shipping and just to complicate things, there were erratically moving fishing boats with their trawls out, both French and British.

The most demanding time was periscope watch-keeping in thick fog. Going round and round on the periscope gazing into a blanket of thick white fog was utterly disorientating. There were no reference points, no way of telling whether one was looking for'ard or aft, other than groping the deck-head for some familiar lump of metal. At periscope depth a submarine is at its most vulnerable but we had no option. We had to snort to run the diesels and re-charge our batteries.

The daily anti-submarine exercises went on until late evening. On completion, most submarine COs would have opted for a quiet night at sea but not Mike Tuohy. He had us hammer into Portland Harbour, arriving near midnight, simply so he could open the bar. As the bar was in the Wardroom where the seats were bunks, that meant none of us could turn-in. We had then to join him in demolishing a bottle of whisky before sailing again in the early hours. At face value, that seemed damned inconsiderate – we needed sleep – but this was his brand of leadership. About halfway down the first bottle, he would switch from idle banter to team talk. That was when we would be informed of our mistakes during the day. It was teambuilding par excellence, though the same couldn't be said about the whisky. From time to time, he would even take a bottle round to a sailors' mess, get them out of their bunks and share it with them. It was highly irregular but the sailors loved it. Apart from the free dram, they had the Captain all to themselves with no other officer present.

On one such evening, Petty Officer Robinson, my radio maintainer, a robust Ulsterman with twenty years of submarine service, was toiling away in freezing conditions on top of the fin, trying to mend our damaged whip aerial. To show that I had not forgotten his endeavours, I popped up several times during the night to see how he was progressing. On my third visit, he looked up from his prone position and said: 'If you want to make yourself useful, Sir, bring me a drink.' It was an excellent tip.

I had helped him recently with a difficult personal problem which required granting compassionate leave. As we sat on top of the fin in the middle of the night savouring our brandies, he said: 'You, Sir, you are the best Divisional Officer I've ever had.' Wow! It was the first compliment I'd had as a submarine officer and it didn't half feel good. Praise from below is every bit as good as praise from above. Coming from a tough old salt like Robinson, it was a real shot in the arm.

CHAPTER 8

Corporate Constipation

*'And o'er and o'er the morning breeze
Came hordes and hordes of shitty fleas.'*

MIKE TUOHY

Following our Portland running, we went west of the Outer Hebrides to participate in a major submarine-v-submarine exercise. In this we would be up against HMS *Valiant*, a nuclear-powered hunter-killer SSN; USS *Nautilus,* the world's first nuclear-powered submarine and HMS *Oracle,* a sister boat to *Otter*. We had no chance against any of them and Mike Tuohy was well aware of that. Worse still, we had developed what was called a shaft rub, a squeak every time the propeller went round. That left us with no hope of sneaking through two weeks of anti-submarine exercise undetected.

As we would be at sea on St Andrew's Night, I proposed holding a St Andrew's Night concert, to be performed via tape recordings played over the internal broadcast system, as there was no assembly space on board. Tuohy thought it a brilliant idea and immediately gave orders for us to snort non-stop until St Andrew's Night to get as far ahead of our pre-planned exercise track as possible so we could then go deep and quiet and enjoy our concert in peace.

If we had been vulnerable before, we were like a matador's cape now. Our diesels would be detected easily as we snorted along. In the far distance, we could hear *Valiant's* mighty sonar rippling away; it sounded like a child running its finger along a keyboard. She was hunting for us, doing what was called 'sanitising the area.' That meant she was travelling at speed whilst sweeping the ocean for miles ahead with her high-powered active sonar. (The one the Soviets copied via the Portland spy ring). Then she went silent.

Half an hour later, the words 'Oscar Oscar Oscar' came echoing out of our underwater telephone. It sounded like someone calling us from the far end of a large cathedral. It was *Valiant*. She had closed the range, was now beneath us and was claiming a kill. It was strange to hear a human voice suddenly speak to us from the deep and even stranger to think that a few hundred feet below us was an aggressive nuclear submarine with a hundred and twenty men on board. It was a breath-taking display of how far submarine warfare had advanced in the Nuclear Age. Had we been at war, we would have been dead meat.

Mike Tuohy took only minor academic interest in the attack; he knew there was little we could do. His priority was to complete the poem he was writing for our concert. He had simply disappeared into his eyrie, the little pressure tight capsule in the fin, and for days we had scarcely seen him. Then, on St Andrew's Night, he ordered us to go deep to stage the concert, a manoeuvre that utterly baffled the post-exercise analysts. They could find no tactical explanation for *Andrew* simply disappearing for twenty-four hours.

The concert was a resounding success. Ordinary Seaman Harrison, one of our youngest sailors and a guitarist, wrote a song for the submarine. The Leading Cook gave us a monologue and I produced a comedy sketch and gave a harmonica recital. The Captain, however, trumped us all with a Chaucerian masterpiece of polite vulgarity.

Alas, his poem is lost to posterity but I do remember the immortal lines: *And o'er and o'er the morning breeze came hordes and hordes of shitty fleas.* Sadly, Mike Tuohy is no longer with us but in his honour, I have written this less earthy parody.

BIOLOGICAL WARFARE

Off Inistrahull they formed their swarm,
A great black cloud like looming storm,
Then o'er and o'er the morning breeze
Came hordes and hordes of shitty fleas.

They'd gathered up in hellish swarm
Obscuring sun in fearful form,
Then out across those Irish seas,
Came hordes and hordes of shitty fleas.

A U-boat, lurking off the Mull,
Observed the sky becoming dull
Then, sucked in by its engine power,
Those fleas flew down its conning tower.

A billion-trillion tiny biters
Hurtled down like midget fighters,
And on arrival did connive
To turn the boat into their hive.

Inside, the boat went black as pitch.
The luckless crew began to itch.
So down beneath the troubled seas,
It dived to dodge those shitty fleas.

Alas, there was no place to hide.
The hordes of fleas were trapped inside
And once they'd settled to the dive,
Began to eat the crew alive.

In the ventilation trunks,
Up their sweaters, in their bunks,
The crew were driven to their knees
By hordes and hordes of shitty fleas.

They breathed them in. They spat them out.
It was a comprehensive rout.
They rubbed themselves with engine grease
To save them from those hungry fleas.

The crew were driven mad as kites.
They had a trillion tiny bites.
The hull resounded to their pleas
For God to rid them of the fleas.

The Captain surfaced in blind panic
And in the situation manic,
The sailors all jumped overboard
To save them from the nibbling horde.

Off Inistrahull is where they drowned.
The submarine was never found,
But somewhere in the seven seas,
Is still being driven by those fleas.

Meanwhile, *Oracle* was having biological problems of a different sort. Her heads (toilets) were blocked and seventy desperate men were dealing with corporate constipation.

In modern society, we have become disconnected from our personal waste. Like inner city children who think milk comes from supermarkets with no concept of cow involvement, we sense no connection between our personal products and the sanitary oblivion of the local sewage works. This cannot be said of submariners. For them, sewage disposal is a major priority. In their infancy, submarines relied on the bucket-and-chuck-it method, which necessitated surfacing. As that was too risky in time of war, the single man sanitary installation was introduced. Consisting of a reinforced lavatory pan with a bolt down lid and a high-pressure air bottle, it was essentially an underwater crapshooter. When a man's offering was safely bolted down, he would open the seacock followed by the air valve and whoompata, his offering would be fired into the deep blue yonder. To render the installation ready for the next user, the man would then have to de-pressurise the pan and re-open the lid, a high-risk operation for if he got it wrong, he got his own back.

Oracle actually had the luxury of a sewage tank but its contents had still to be blown overboard every other day using high-pressure air. When the pressurised air in the emptied tank was then vented inboard, the whole submarine had to endure what can only be described as a corporate fart, an odour most foul that persisted until filtered-out by seventy pairs of human lungs.

In time of war, an alert and inquisitive enemy might wonder at the source of human faeces popping up from the deep or even just an unexplained brown patch in the ocean. He may even be moved to drop a clutch of depth charges on the spot to satisfy his curiosity. Thus, one badly timed excretion could prove fatal for the submarine. Sewage control is a significant military factor and the human bowel is at its most active in the heat of battle. A spy hiding in a cupboard, for example, dare not reveal his presence with a smelly indiscretion. In submarines, the discharge of sewage is a Command decision.

The uninformed citizen may imagine that men in submarines worry about sinking or the morality of launching nuclear weapons. Nothing could be further from the truth. Their main concerns are food, sewage and movies. Consider the case of a submarine with a crew of one hundred and twenty men. The design calculation would assume one daily bowel movement per capita. If each man were to be credited with a British Standard three-pounder plus a bucket of flushing water, the sewage tank would be filled in two days. Tank contents would then have to be discharged through the sewage overboard hull valve, in biological terms the submarine's anal orifice. Were this to become blocked, the immediate consequence would be the aforementioned corporate constipation.

In *Oracle's* case, the blockage could not be removed and the desperate but ingenious crew had resorted to the use of waste paper baskets as impromptu lavatory pans, with large brown 'On Her Majesty's Service' envelopes being used to catch the product. These were then delivered by hand to a large biodegradable underwater bin bag that was fired from the Gash Gun, a small downward pointing torpedo tube. It is small wonder that submarines are known in the trade as sewage tubes.

Dear Reader, next time you lock yourself in the wee room underneath the stairs, count your blessings. It is most unlikely that a passing aircraft will bomb you; it is equally improbable that the public sewage system will backfire on you; and it is inconceivable that if your cistern overflows, your house will fill with water and sink. So spare a thought for the poor, bloody submariners.

At the end of the exercise, we returned to Faslane where I was re-united with Kate and the family. She was now eight-and-a-half months pregnant. Recognising this, Mike Tuohy kindly allowed me to have a couple of extra days at home and re-join the boat in Campbeltown. (In those days, paternity leave was unimaginable). From Campbeltown, we sailed round the North of Scotland to Esbjerg, the principal fishing port of Denmark. On our first night in, I received a telegram from Kate. She had just successfully delivered our second son. We called him Andrew, of course. Immediately, I grabbed two bottles of Glenfiddich from the Wardroom, mustered the Duty Watch in the Control Room, and we wet the baby's head.

****

A few weeks later, *Auriga* suffered a catastrophic main battery explosion. Her forward accommodation space was completely wrecked, the deck having buckled upwards shattering all the woodwork and bending the steel stanchions. All power on board was lost, meaning the submarine was completely disabled and soon was filled with toxic smoke. Several of the crew who had been asleep in their bunks above the battery compartment were badly injured but thanks to the bravery of their shipmates, were rescued and survived. Fortunately, the crippled submarine managed to limp back to Plymouth.

Charging main batteries is a regular daily routine in diesel submarines and as batteries under charge give off large quantities of hydrogen, an effective hydrogen clearance system is vital. In *Auriga*, a flap-valve had fallen shut allowing an explosive level of hydrogen to build up. It then took only one spark to cause an explosion, a sobering reminder of the many life-threatening issues one has to manage in a submarine.

Our next sortie was to Bayonne in France accompanied by *Olympus*. We were to rendezvous off the mouth of the River Adour, which begins life at Col du Tourmalet in the Pyrenees, well known to followers of the Tour de France cycle race. Typically, Mike Tuohy decided that we should sneak up to the rendezvous position dived. As I was on periscope watch, I was first to sight *Olympus* sitting on the surface waiting for us.

I pressed the transmit button on the periscope microphone and called her in French on VHF. 'Sous-marin Anglais. Sous-marin Anglais. Ici Bayonne. Ici Bayonne. Over.'

Through the high-powered periscope I could see her Officer of the Watch flapping about on the bridge but there was no reply. I called again. 'Sous-marin Anglais. Sous-marin Anglais. Ici Bayonne. Ici Bayonne. Répondez s'il vous plait. Over.'

I could almost hear him shouting down the voicepipe: 'Tell the Captain the French are calling us in bloody French.' With the remarkable instinct of a linguistically challenged Brit, he called back: 'Bayonne, wait.'

By now, Mike was on the Attack periscope convulsed with laughter.

I repeated my call with a tone of greater urgency. 'Sous-marin Anglais. Sous-marin Anglais. Ici Bayonne. Ici Bayonne. Bonjour. Over.'

After a short delay, a voice called back: 'Bonjour Bayonne. Ici sous-marin *Olympus*. Parlez vous Anglais?'

Mike recognised the voice as that of the Captain and decided it was time to make our entrance. 'Surface in emergency,' he ordered.

Every ballast tank was blown and we smashed through the surface like a leviathan rising from the deep, only two hundred metres short of *Olympus*.

'Bonjour *Olympus*. Ici *Andrew*. Good morning,' Tuohy called in his most theatrically pukka voice.

For the visit, we carried two Guards Officers, Jimmy and Jamie – Jimmy from the Coldstreams and Jamie from the Scots. They were to lay wreaths in the Guards

Cemetery at Bayonne, a legacy of the Spanish Peninsular War. They were entirely different characters but confirmed every Naval Officer's impression of Guards Officers. We served together in Her Majesty's Forces but otherwise lived in completely different worlds.

Jimmy, the sophisticate, was Adjutant to the Honourable Artillery Company, the oldest regiment in the British Army. It was also the City of London Regiment, drawing its volunteers from the ranks of City stockbrokers and bankers. Jimmy, a regular soldier, had volunteered to be its Adjutant as he preferred London to garrison duty in the British Army of the Rhine or Northern Ireland. He had the louche, languid style of a penniless lord of the manor, the cuffs of his old blazer being frayed and the elbows threadbare. He also talked with the aristocratic presumption of superior knowledge. I classified him immediately as a bullshitter trading on the principle of the one-eyed man in the land of the blind. He owned a vineyard near Bayonne, the ulterior motive for his visit, and was planning to start a business taking rich Americans on beer-tasting tours of Europe. I struggled to imagine him being an expert on European beer and suspected that he thought the Americans would be an easy touch. Men of Jimmy's ilk are so replete with self-confidence that ignorance is never a handicap.

Jamie was a different kettle of fish. He seemed to have been born with neither intellect nor sense of humour. He was pure infantry, born to be a soldier. I struggled to hold a meaningful conversation with him but had no doubt that he would have been utterly fearless in trench warfare. Jamie, however, was of more practical value than Jimmy. He had brought his piper with him.

And so we arrived in Bayonne with a Scots Guards' piper playing on our fin. The French loved that and so did Tuohy. We completely upstaged *Olympus*.

The Captain was in his element in Bayonne; not only was it Basque country but also rugby country. By the time our combined rugby team was being thrashed by the locals, Tuohy had acquired a Basque beret and a makila, the large spiked stave used by Basque shepherds in the crags of the Pyrenees, and was strolling around the touchline like a Basque grandee. Mike Tuohy had style.

That evening, I attended a formal reception hosted by the British consul and his wife. As Consul is an honorary position, both were French. During the reception, the normally loose ensemble of a cocktail party suddenly began to form a circle. Had we been in a Glasgow pub, it would have been a sure sign of a fight in the offing. It was, but not with fisticuffs. In the centre of the ring stood Jimmy facing the Consul's wife, each holding a *taste-vin,* the shallow cup used for professional wine tasting. Between them on the floor, someone had placed a spittoon.

Apparently, the bold Jimmy had contradicted the Consul's wife on the subject of French wines. Unfortunately for him, Madame Consul happened to be the daughter of a *viticulteur* (wine producer) and recognising British bullshit, had challenged him to a wine-tasting competition.

A hush descended as the referee served the first sample of anonymous wine. Both contestants sipped it, swilled it round their palates and spat into the spittoon.

'It's certainly a Bordeaux,' Jimmy pronounced. 'Probably *Saint Emilion* but I'm not sure of the vintage. Could be 1950. That was a good year.'

Then it was Madame's turn. 'Non. Thees ees from furzer north, a Beaujolais, thees year's crop, I theenk.' She pouted her verdict in a beguiling French accent and was spot on.

One up to the French – but at least Jimmy had the colour right; it was red.

The colour was switched to white.

'It's a Chablis,' proclaimed Jimmy.

'Non. Eet ees a Muscadet,' countered the Consul's wife. Two-nothing to the French.

At five-nothing, the Consul's wife took pity. 'Alors,' she said, turning to the crowd. 'I must compleement Monsieur Jeemmee. He knows zee names of many French wines.'

****

It was a joy to serve under Mike Tuohy. He was inspirational, a brilliant Commanding Officer with an insatiable appetite for mischief. I would have happily gone to war with him and that is the greatest compliment one submariner can pay another. I shared his natural sense of irreverence but I always make sure I am on firm ground before sticking my neck out, whereas he had utter contempt for authority and that did him no favours. He even took his submarine to sea one Sunday afternoon without permission with only the duty watch for crew, simply because he disapproved of the berth he had been given. The High Command was not amused. He was the only one of my Commanding Officers not to be promoted beyond Lieutenant Commander, the rank he held in *Andrew,* but for me he was a role model.

In submarine terms, Mike Tuohy was the last of the Mohicans. There was no place for such a character in the nuclear navy. The question was: was there a place for me? I had been heading backwards into the old Navy with its fabulous characters.

CHAPTER 9

God of the Underworld

'Resurgam'
(I shall rise again)

MOTTO OF HMS OSIRIS

From *Andrew*, I was appointed to *Osiris*, a 'special fit' boat, i.e. fitted with special surveillance equipment. Her motto, *I shall rise again,* seemed apposite; my career had been resurrected. In ancient Egyptian mythology, *Osiris,* brother of Isis, was God of the Underworld and normally depicted as having green skin, another good omen as Electrical Officers were known as 'Greenies' on account of the green cloth originally worn between their gold stripes. Osiris was also the God of Silence, the one to pray to on Sunday afternoons when your neighbours flash up their power-driven lawnmowers. By all accounts, he was the ideal god for a submarine Electrical Officer.

In *Otter*, the joke had been that *Osiris* didn't have a bar. That suited me but was untrue. She even had a brandy tank built into a Wardroom cupboard. The difference was that she had been employed on spying missions and that demanded much more work in harbour in terms of writing reports, collating records and liaising with the Intelligence community. She was a serious boat.

By the greatest of good fortune, the Marine Engineer was John Holl, one of the best buddies a man could ever wish to have in a submarine. He had been five terms ahead of me in my Division at Dartmouth and was in my year at Manadon. Not only that but he had been a fellow member of the Cross Country running team and the Quart Club (an Engineers' drinking club), as well as being trombonist in the College jazz band. For the first time, I had a professional soul mate.

John had gone straight from Manadon into the nuclear world and had just arrived from *Warspite* in which he had been Assistant to (Admiral) Patrick Middleton, one of the Navy's most experienced nuclear engineers. He had also been under the Command of (Admiral Sir) 'Sandy' Woodward who later rose to fame as our victorious Admiral in the Falklands War of 1982. *Warspite* had been engaged in close-range surveillance of the Soviet Northern Fleet; so close that she had suffered an underwater collision with a Soviet *Echo 2* cruise-missile-carrying nuclear submarine that she had been trailing – reported as 'collision with an iceberg.' The damage to her fin had been so serious that she could not return home until a team of shipwrights from Faslane had joined her in a remote Scottish loch to camouflage the damage. John opened my eyes to the very different culture of nuclear submarines.

Alex Gadsby, the First Lieutenant, known in the Flotilla as 'Gloria', was a decent sort but went pop-eyed when angry. David Girvan, the Sonar Officer, was competent but strangely unpopular. It was he who had borrowed Ian Tallant's Jaguar without permission and left it burnt out at the roadside. He had not confessed to that crime but when the evidence pointed to him, his father had simply paid for a new Jaguar. His father, it seemed, was a rich banker based in Switzerland with a house in exclusive Eaton Square for use when he was in London. According to David, his father had divorced his mother and married a glamorous young woman of David's own age and he even confessed to be lusting after his stepmother. One day, he arrived on the jetty in a chauffeur-driven Rolls Royce belonging to his father, far too ostentatious for my liking.

Willie Woodard, the Navigator, was the son of an Anglican vicar and had a charismatic older brother, a Fleet Air Arm pilot who became Commodore Clyde and then Flag Officer Royal Yacht. Willie was, to say the least, relaxed. In fact, he took the art of nonchalance to a new level but had such charm that he got away with murder. Rather than meticulously calculating the required speed for reaching a destination on time, he simply planned all his passages at full speed. That, he said, saved him the effort of doing calculations that he may get wrong. And so, under Willie's navigational regime, we always arrived ahead of schedule – and then had to kill time just short of our destination. Thus he ensured that he never made the greater mistake of arriving late, but it knocked the hell out of my generators. On the long run back from Gibraltar, we actually arrived in the Clyde the day before and had to spend a night going up and down in the estuary at slow speed when I could have been at home with Kate who was only twenty miles away.

Years later, while reading a newspaper in the departure lounge at Heathrow, I recognised a pair of uniquely dilapidated suede shoes. They were unmistakeably Woodard's. I lowered my paper and sure enough, it was Willie, large as life, and still wearing the shoes he had worn in *Osiris*. It was a joy to meet him again. By then he had left the Navy and was skippering some sort of coaster that plied its trade between Libya and Southern Ireland. He didn't say so but I suspect he was working for MI6 and tracking illegal arms shipments from the Gaddafi regime to the IRA. Either that or he was gun running.

When I joined *Osiris*, Lieutenant Commander Edmund Shackleton Jeremy Larken (later Admiral and Distinguished Service Order) was in Command. He had the same lanky physique as Mike Tuohy but was straight out of the Dartmouth mould, an absolute stickler for detail and for maintaining the highest standards of the Royal Navy. Like Tuohy, he also had a warm, wholehearted personality with some endearing foibles. When under stress, he would develop a stammer, which sounded like crashing the gears in a racing gear change. He had recently acquired the latest in male fashion accessories – a James-Bond-style, black, plastic attaché case with combination locks, ideal for carrying classified papers, but he had the unfortunate habit of opening it upside down. To help him with this problem, Gloria's predecessor had stuck a tally on its upper lid with the word TOP engraved on it. Jeremy would later command *Valiant* and win fame in the Falklands War as Captain of *Fearless*, one of our main amphibious assault ships in the audacious landings at San Carlos, before being promoted to Admiral.

In *Osiris*, I went for my second visit to Esbjerg, accompanied by *Otter*. This time, the atmosphere of fun that had so enriched my previous visit in *Andrew* was missing but Jeremy had organised a piper, Piper Pleasance from the Black Watch, to come with us. He not only played on the fin as we entered harbour but also spent the next three days acting as the Pied Piper of Esbjerg, taking his bagpipes ashore every night and leading an ever increasing crocodile of Danes and submariners from pub to pub.

On our last night in, I was Duty Officer and distinctly remember Mickey Perks, a young Mechanic, coming through the Control Room laughing and chaffing and making rude gestures with a sausage. I was one of the last people to see him alive. The following morning, he missed the boat on sailing. Six weeks later, his body was dragged from the harbour.

At the subsequent inquiry, our trot sentry reported that Perks had gone across to *Otter* to visit his 'oppos.' As they lived in the tail end of the boat, one explanation was that he had left *Otter* by her stern hatch where there was no plank, had tried to jump across and had fallen between the two boats. No one knows the truth but *Otter*, for some inexplicable reason, always seemed to attract trouble. During the Coroner's Inquest in Birmingham, I met Perks' grieving parents. That was the moment I realised that a sailor is also a son. It was difficult to know what to say to them. Jeremy maintained contact with Perks' Mum for many years after – a submarine crew is family.

On passage home from Esbjerg, the weather was foul. By the time we reached the Pentland Firth, it was Storm Force Eleven, one stop short of Hurricane Force. As it was far too dangerous to have the Conning Tower open, we shut down and kept watch through the periscope. We were, in effect, dived but on the surface. Fishermen have no such luxury. I caught sight of a trawler. One minute I was looking up at her, the next she had disappeared completely into a trough between thirty-foot waves. Fishermen have my fullest admiration. I remember that incident every time I buy fish.

By the time we had rounded Cape Wrath and were heading south towards Skye, the Captain decided to take the more sheltered inner route and go through the Sound of Sleat between Skye and the Mainland. This meant going through the narrows at Kyle Rhea where the tide can flow at as much as eleven knots. Jeremy further decided that our piper should be up on the fin to play us through the narrows. A British submarine slipping silently through that remote Kyle with a piper playing in full Highland regalia on its fin was certainly something to stir the blood but the only witnesses were a handful of sheep, the odd deer and some passengers in the tiny Kyle Rhea ferry which diverted for a closer look.

****

Mike Harris (heir presumptive to Baron Harris) was still a Lieutenant when he took Command of *Osiris*. He was another first eleven officer who would go on to become an Admiral. The Submarine Service, being the most powerful arm of the Navy, had now become a well-worn track for officers heading to the highest ranks. I was so very fortunate to be rubbing noses with this calibre of officer in the close confines of diesel submarines whilst still in the infancy of my own career. It taught me to think like a Commanding Officer.

Mike was a very different character from Tuohy and Larken. He was taciturn, appeared shy and seemed uncomfortable imposing his orders on others. He always thought before he spoke and used his black, bushy eyebrows like signal flags. Having said that, he was thoroughly good-humoured, never lost his cool, and was very approachable. As I was now full of confidence and on top of my game, the change of Captain was a seamless transition. Nevertheless, new relationships have to be tested.

On overnight passage down the Irish Sea, I had First Watch on the bridge (2000–2359) and so left the others to enjoy dinner, went into the passageway, donned my foul weather gear and then, for a bit of devilment, returned to the Wardroom curtain. I asked John Holl to go into my drawer and pass me my specs, despite my never wearing them at sea and the bridge having binoculars and a lookout.

Mike Harris arched his eyebrows. 'Do you need specs, Electrical Officer?'

'Yes, Sir.'

His eyebrows arched further. 'Are you fit to keep bridge watches?'

'Not at all, Sir. I failed the eye test for bridge watch-keeping. That's why I'm an Engineer.' He looked stunned. 'But it's not a problem. If I see anything, it's bound to be so close that I always call the Captain.' With that I disappeared up the conning tower.

An hour later, the sea ahead of me was a mass of bobbing lights. It was a fishing fleet in action. There were dozens of fishing boats all doing their own thing, each with a different set of coloured lights to indicate whether it was trawling or drifting, going left or right, working as part of a pair or simply shining searchlights on its incoming nets. It was my job to interpret their lights and avoid colliding with them or running over their nets. As it was standard practice to call the Captain if any ship was likely to pass within two cables (400 metres), I had to call him umpteen times. Eventually, I called down the voice pipe for 'Galleyboots' Kerr, our Leading Chef, to come and speak with me.

'Do you have any unfrozen fish?' I asked.

'Yes, Surr,' he replied, in his thick Glasgow accent. 'We're preparing tomorrow's lunch. It's fush and chips.'

'Would you please give one to the Messenger?'

'Nae problem, Surr.'

When the Messenger had been equipped with his raw fish, I instructed him to take it to the Captain with the message: *With the compliments of the last fishing boat.*

Then we went to the fringe of the Arctic ice cap in support of *Valiant* who was going in under the ice, our role being to provide a radio link. We were also to do a short sortie under the ice ourselves to test an experimental upward looking television camera. The thought of going under the ice in a diesel submarine worried me. A nuclear boat has no need for air but a diesel one is like a whale. It has to come up for air to run its diesels and re-charge its batteries. We could survive for only a day or so without air, and less if we were moving about. I thought of the recent case of two Arctic Grey whales trapped under the ice off Canada. They had managed to find a small hole, just enough to stick their blowholes through and gasp for air. The open sea was too far away for them to reach it on one lungful and that little hole had become their only hope of survival. If it froze over, they would drown. A diesel submarine trapped under the ice would be in the same situation. It was my job to do the battery endurance calculations.

From the ice cap we returned to the Clyde for more Perisher running, returning each evening to a buoy in Rothesay Bay. There, I managed to take my father to sea for a day. Not having been a submariner, he was mesmerised by the drama of us attacking frigates and destroyers which were trying to run us down at high speed. During the war, he had seen submarine warfare from the other end of the telescope. Having seen ships explode and men die at the hands of German U-boats, he had no reason to like submariners but his pride in me was tangible.

During that Perisher, we conducted my first practice firing of a *Mk23* homing torpedo, the weapon we would use against Soviet submarines in time of war. The beast had one simple listening device on its nose, was hard of hearing, had zero intelligence and would happily attack a snapping shrimp if it heard that first. It trailed a control wire with which the submarine could guide it to its target, but guiding a torpedo against a submarine at an unknown depth was like trying to pin the tail on the donkey blindfolded. Its ineffectiveness came as a shock. In the Submarine School I had been taught how it worked, but not how inadequate it was. The reality was that we had enormously expensive, hunter-killer, nuclear-powered submarines that were brilliant at hunting but little more than nuclear-powered water pistols when it came to killing. We would be deluding ourselves to imagine that if went to war with the Soviets, our much vaunted SSNs would achieve a significant kill rate against Soviet submarines. A new *Mk24* torpedo was under development but seemed to have disappeared into the fog of poor project management and was on the brink of cancellation.

There was nothing we submariners could do about this except train to fight with the weapons we had and we were certainly doing that with the utmost professionalism. However, the breath-taking success of our nuclear submarines in intelligence gathering was, I feared, creating a false sense of our capability. Had it come to a shooting war, much of our advantage would have been lost through the inadequacy of our torpedoes.

Our other torpedo, the *Mark 8*, was simply a relic of the Second World War or, more accurately, the First World War. It was a diesel-driven bomb that ran in a straight line and was fired in salvoes to allow for some missing. It was designed for periscope attacks against surface ships, had no homing system and was useless against dived submarines.

It was obvious to me that we desperately needed an effective new anti-submarine torpedo and a new breed of Weapons Engineer dedicated to the development of underwater warfare. Here I saw a career opportunity. I was now so far behind my contemporaries who had advanced into the world of nuclear reactors and Polaris missiles that I had no hope of catching up and so underwater warfare was where I would pitch my efforts. I imagined myself as the first of a new breed and wrote a paper on the subject, which I presented to Teacher, the wonderfully avuncular Commander Frank Grenier who would later become FOSM, the Admiral in charge of the Submarine Service. It was very well received – credibility at last.

When he became Flag Officer Submarines, I wrote this poem in his honour:

> Great good and generous Admiral Grenier,
> There is no finer Admiral anywhere.
> Bold warrior of the North Atlantic
> Always calm, never frantic.
> From Gosport up to Tobermory,
> Thine is the power, thine the glory.
> On Earth, Thy will be done
> By SMs 10, 3, 2, and 1. (our four submarine squadrons)

****

Our next excitement was a towed array trial. The towed array was a top-secret experimental kit that promised to (and did) revolutionise underwater warfare. The array was a kilometre long with a collection of listening devices at its far end. As we had no towing hook, it had to be wrapped four times round the raised Attack Periscope to take the strain. The breakthrough it promised was that it allowed our listening hydrophones to be removed from the submarine's hull and thus escape being deafened by the submarine's own machinery, a major problem in sonar.

Associated with the towed array was a thing called a 'narrowband analyser,' a product of the new digital technology and some very clever mathematics. In simple terms, it would enable us to pluck out man-made machinery noise from the cacophony of natural sea noise, something the human brain cannot do. The sea is very far from silent; it sounds like a rookery at dawn or the jungle by night. With the new kit, it would be like sitting in a concert hall listening to a symphony orchestra but still being able to identify the hum of the ventilation fans. At least that was the theory. We now had to prove it.

The practical problem was that the array had twenty-six individual wires that had to be passed through the pressure hull and we had no such through-hull connection gland. There was therefore the need for a twenty-six-pin plug and socket outside the pressure hull to allow the array to be connected. The Admiralty scientists' solution was to divide the wires into two groups and pass them as separate bundles through our emergency navigation light glands, thus creating a Y-fork electrical junction outside the hull.

The trial took place off St Kilda to the west of the Outer Hebrides and was almost a complete failure. Every time we dived, the plug and socket arrangement flooded, put a full earth fault on our electrical system, and forced us to surface. We then had to recover a kilometre of wire by hand and bob about on the Atlantic swell while the scientists spent hours fitting new watertight plugs and sockets. And when we dived again, the new plugs and sockets would flood immediately. After the third failure, an exasperated Mike Harris sent for me and said: 'Electrical Officer, do something. We're running out of time.'

As it was obvious that the watertight plugs and sockets were not watertight, my solution was to get rid of them altogether and make twenty-six individual wire-to-wire watertight connections. So I took over responsibility from the scientists and called for their Y-fork connection and a working length of array cable to be hauled down the Conning Tower into the Control Room where the wires could be cut, dried and individually re-connected in watertight sleeves. For good measure, we then put the bundle of watertight connections inside a length of radiator hose plugged at both ends with wax sealant and filled with engine oil. Heath Robinson would have been proud of me.

It was a lengthy procedure, about four hours in all, with my Chief Electrical Artificer sitting cross-legged on the Control Room deck working away with a hair dryer and soldering iron. *Onyx*, who had come with us to act as our target, had now been lolling about uselessly for two days and the crew were mightily bored. We had already sent her six bottles of Champagne by heaving line as a peace offering. As time went by, Mike Harris became more and more agitated and kept sending for me to ask how much longer we would be. We had less than twelve hours left to complete the trial.

Eventually my team finished the job and I reported proudly that we were: 'Ready to go.'

Immediately Mike ordered: 'Diving Stations.'

As he came striding into the Control Room, I glanced up at the conning tower ladder and my heart sank. The main cable of our electrical spaghetti had been passed through a rung of the ladder. Now we could neither lift the cable back up the conning tower nor shut the conning tower lid. We were unable to dive. I had crippled the submarine. We would now have to unpick twenty-six watertight joints wire by wire, untangle the wires from the ladder, and start all over again.

John Holl then appeared at my shoulder. As ladders are mechanical, the ladder was his responsibility. 'Don't worry, Eric,' he said with a broad grin. 'We'll saw it in two.' And we did. Situation saved.

Now we could get the array cable up the conning tower and over the side but the Officer of the Watch and lookout had to stand on other men's shoulders to get up and down from the bridge. When we dived, all our watertight joints held and the trial proved to be an overwhelming success with far-reaching consequences. For this triumph, John's department presented me with a framed 'Cock of the Month' certificate but the grateful scientists hailed me as a hero and put our handiwork on display in their museum at the Research Laboratory in Teddington.

The success of that trial changed the face of underwater warfare. We would now be able to detect Soviet submarines at very much longer ranges. It would give us a massive detection advantage over them. In future, towed arrays would be fitted to all our submarines and to anti-submarine surface ships as well. The latter were able to deploy their arrays at depth, thus avoiding the perennial handicap of surface sea noise in rough weather. However, *Osiris* remained with only half a conning tower ladder until we got back to Faslane where the two halves were welded back together again. With the top half of the ladder ending at head-height, our bridge watch-keepers developed considerable gymnastic skills.

During that three-day trial we had lolled about on the surface within sight of the spectacular islands of St Kilda, the jagged remnants of an ancient volcano that had risen from the sea aeons before. The islands lie forty miles (64 Km) west of the Outer Hebrides and are little more than rocky outcrops fully exposed to the worst weather the Atlantic can provide. It was difficult to believe that human beings had actually lived there but since prehistoric times they had. Their staple diet was seabird, mainly gannet, with the odd potato and a bit of barley thrown in. Sheep had provided them with milk, cheese, wool and, I presume, the odd leg of mutton. The population was never more than one hundred and eighty and they had lived as a commune. By the time the islanders were finally evacuated in 1930, the population had reduced to an unsustainable thirty. The only human residents since then had been military or National Trust for Scotland personnel, with one of the islands, Hirta, now hosting a radar station and the archipelago being a World Heritage Site.

But we were not the first submarine to have viewed St Kilda from such close range. During the First World War, a German U-boat had come into the bay and destroyed the Admiralty wireless station on Hirta with its gun. The islanders declared it, in Gaelic of course, to be a good submarine as it made no attempt to shell their little string of hovels.

From the distance, the islands looked white, as if covered in snow. They were not. They were covered in guano, or the producers of the guano, the magnificent Atlantic Gannets. With an estimated fifty thousand mating pairs, St Kilda hosts the world's largest colony of the species. When our trial was complete we set off on the surface for a visit to Alesund in Norway and, as if to see us off its premises, a gannet took station about five-feet above my head while I was on watch. It was riding the up current of air from the front of our fin like a surfer riding a large wave. For about an hour, this magnificent creature kept me company, gliding in total silence above my head with barely a twitch of its wings. It was a cold, clear, summer morning and the bird was utterly relaxed. It was simply enjoying the luxury of a free ride. It was a life-enriching experience. I could almost have reached up and touched him. For that wonderful hour, it felt as if we were flying together.

In terms of pure elegance, the gannet is one of Nature's aristocrats. It has long, slim, black-tipped wings with a span of almost six feet (180 cm). It is built to glide. He or she, I'll call him 'he', was pristine white with a distinctive yellow head, piercing blue eyes, and a long, grey beak with thin stripes reaching back to his eyes. He was not remotely like the seagulls one sees following a ferry in search of food or fighting over discarded fish-and-chip packets on the promenade. The gannet seeks nothing from humans – other than for us to leave some fish in the sea.

When a gannet feels hungry, it soars up to ninety feet (27 m), looks for a shoal of fish and comes plummeting down like a bullet at sixty miles an hour (100 kph), folding its wings on the way down before hitting the water so hard that his neck and forehead have been reinforced by Nature. A gannet can dive down as deep as a hundred feet to snatch an unsuspecting fish and will happily go off on a three-day fishing trip of up to three hundred miles (480 km), before returning to his wife and kids. Gannets mate for life and when the mate returns to the nest, the reunion is celebrated with a joyous clashing of necks and beaks, the gannet equivalent of a passionate embrace.

And so I said farewell to St Kilda. It had been a week of amazing contrasts; two submarines engaged in top secret trials with state-of-the-art technology, Heath-Robinson ingenuity, the spectacular remnants of a prehistoric undersea volcano, the purity of the open ocean, and the close company of one of Nature's most graceful creatures. (I refer not to our First Lieutenant).

****

On return to the Clyde, we were tasked with investigating the Soviet spy trawler that kept permanent station between Ireland and the Mull of Kintyre throughout the

Cold War, the 'Malin Head AGI' as it was known in NATO-speak. Its role was to monitor communications and submarine movements in and out of the Clyde, both British and American. We also assumed it was relaying our submarine movements to a Soviet submarine waiting in the wings to the west of Ireland, as the Soviets were attempting to track our Polaris submarines leaving the Clyde. The spy trawler could never have been mistaken for an innocent fishing boat. It was festooned with aerials and had no fishing gear. We could see what it had above the waterline. The big question was: what did it have underneath? Our job was to find out.

As we rounded the Mull of Kintyre on the surface, I sighted it sitting in its usual position, drifting on the tide. It would sit there for months on end and I wondered what on earth its crew did to pass the time. I imagined they were revelling in the joys of British or Irish television, which must have been vastly more entertaining than the propaganda-riddled, State-controlled, Soviet equivalent. Our plan was to ascertain its exact location, head out to sea, dive, circle back, and do a covert under-run. When it saw us, it got underway and headed towards. I duly informed Mike Harris of its approach.

It passed close by on a reciprocal course down our starboard side, did a U-turn under our stern and came up parallel with us a hundred metres off the port side. I could examine it in minute detail. It had a very much larger crew than was normal for a trawler of its size and most of them seemed to have come up on deck to see us. The number of women crewmembers was a surprise but I saw no gorgeous, blonde, Julie-Christie-like Lara, as in Doctor Zhivago. These women looked like navvies and wore overalls with industrial aprons. It was strange to see 'the enemy' at such close quarters. They were fellow human beings and looked a pretty poor lot. I felt sorry for them and wondered at the low quality of their lives in a Communist-controlled state.

A man then came out on their bridge wing and trained a loudspeaker at us. Within minutes, it was pumping out strict tempo dance music. How amusing. We were in the Beatles-Elvis Presley-Rolling Stones rock 'n roll era and they were playing the dance music of my parents' generation. Was this a joke or were they trying to impress us with Soviet cool? Were they trying to establish a rapport with fellow seafarers or was it a poorly calculated act of Soviet propaganda? I'd like to think the former.

Mike Harris arrived on the bridge. 'Ignore them,' he said. 'Look straight ahead. If we were loitering off Archangel, I'd wave to them but not here.'

The Captain, the lookout and I, three steely-eyed submariners, all wearing uniform navy-blue berets and white roll-neck submarine sweaters, were staring grim-faced straight ahead while within hailing distance, a Soviet spy trawler was bombarding us with dance music. We increased speed and pulled away but as soon as we were over the horizon, we dived and returned, this time passing directly underneath the AGI and photographing what she was hiding under her skirts. It turned out that she had a torpedo tube! That was a surprise. Legally speaking, it made her a warship.

As we had passed a mere two metres beneath her hull, she must have detected us for she increased speed and by the time we had returned for a second look, we had entered into a high-speed underwater chase.

****

Our next assignment was the NATO Perisher's Ocean Phase, a major submarine-v-submarine exercise to the west of Ireland, not the high intensity ducking and weaving of an attack against fast-moving frigates. This was about listening, detecting, approaching and launching simulated attacks against other submarines. During the exercise, whilst on periscope watch, I caught sight of something orange in the water; it looked like a lifejacket. As the sea was rough and the periscope only a few inches above its surface, something small like that could be seen only in glimpses when it happened to be on the crest of a wave. As an aircraft had ditched recently somewhere to the west of Ireland, I called the Duty Captain, a Dutch trainee. He and Teacher came to the Control Room immediately and manned both periscopes.

A few moments later, the Dutchman shouted: 'I haff it.'

'Where?' Teacher called.

'Ninety degrees to starboard.'

'What is it?'

The Dutchman hesitated. 'It's a…teddy bear.'

'Well rescue it!'

'Standby to surface to rescue teddy bear,' the Dutchman ordered over full main broadcast.

We had no drill for rescuing teddy bears but we surfaced in emergency and slipped immediately into full Man Overboard drill.

Rescuing a man overboard from a submarine at sea is dangerous and difficult. In the prevailing conditions with waves breaking over our casing, men would be washed away like chaff in the wind if they were not hooked on to the safety rail that ran along the deck. The rescue itself would require the Ship's Swimmer in full rubber diving suit to go into the water with a safety line attached while two other men stood by clipped to the safety rail, ready to haul both him and the casualty back on board. It might only have been a teddy bear but rescuing it was just as dangerous as rescuing a man. As Photographic Officer, I went up into the fin to capture shots of the rescue through the fin door.

When I suggested to Mike Harris that the rescue might be the sort of light-hearted titbit the BBC liked to tag on to the end of the national News, we sent the following press release: *HM Submarine Osiris breaks off major NATO exercise to rescue teddy bear. Photographs on return.* Three weeks later with the incident all but forgotten, we returned to Faslane to an unexpected hero's welcome. There was a posse of journalists waiting for us, all clamouring for copies of my photographs.

The bear itself remained a mystery. On its bottom it said *Made in Taiwan*, not much of a clue. Where had it come from? Had it fallen off a passing ship or been carried all the way across the Atlantic on the Gulf Stream from the Caribbean? As we had been well to the west of Ireland, it was unlikely to have drifted out from the UK. If only we could have found its owner and returned it, now that would have been a story.

CHAPTER 10

Walter Mitty

'Beautiful things don't ask for attention.'

JAMES THURBER

One idyllic evening, whilst on passage back into Gibraltar from a Mediterranean-based Perisher, an OFFICERS DELTEX signal was received, DELTEX being the acronym for Delicate Text. It meant bad news of a personal nature for one of the ship's officers. As Officer of the Watch, I was informed of its receipt and it sent a shiver up my spine. Was it about me? Had something happened to Kate or the boys?

When we were back alongside, Gloria, John Holl, and David Girvan disappeared into HMS *Rooke*, the naval barracks. Why had I been excluded? I was the third head of department, not David. Was it about me? As our main generators required attention, I nursed my anxiety and went back aft to discuss repairs with my team.

By the time I came forward again, Willie Woodard was the only other officer on board. He was sitting alone in the Wardroom mustering our Confidential Books. As this was abnormal for Willie, who was years my junior, I assumed that he too was in on the secret and my dudgeon rose to new heights. I sat in silence for a moment and sipped a horse's neck (brandy and dry ginger). Eventually, Willie looked up from behind his pile of heavyweight books that ranged from Confidential to Top Secret, all designed to sink if thrown overboard, including our entire holding of secret cryptographic codes. 'Do you know what's happening?' he asked.

'No, haven't a clue.'

'Neither have I. I thought you would know. Gloria just told me to muster all the Confidential Books. That's David's bloody job, not mine.'

It was all very mysterious.

The following morning before sailing, Mike Harris called all the officers to the Wardroom for a briefing.

'Gentlemen,' he said. 'Yesterday evening, I received an OFFICERS DELTEX from the Squadron. There have been a number of thefts recently from officers' cabins at Faslane, including some suits. As the prime suspect is Lieutenant Girvan,

I despatched him under the escort of the First Lieutenant and Lieutenant Holl to search his cabin in *Rooke*. As some of the stolen items were found there, he has been arrested and is being sent back to Faslane under Provost Marshal (naval police) escort for questioning. So we shall be an officer short.'

David never made it to Faslane. He slipped his escort in London and disappeared. That provoked further signals to *Osiris* seeking all known details about him. These unearthed some astonishing facts. His father was not a filthy rich banker living in Switzerland with a glamorous young second wife. He was a Scottish schoolteacher living in the Borders with his first wife. Ian Tallant's Jaguar had been replaced at their expense. The chauffeur driven Rolls Royce that had delivered David to the jetty at Faslane had not been his father's but hired by David to impress us. The man was a real life Walter Mitty – and a thief.

Further investigation revealed that he had also embezzled the Tobacco Fund for which he was accountable and he had failed to insert the necessary amendments in our Confidential Books for which he was responsible. Worse than that, he had falsely certified that the amendments had been inserted, which could have had serious military consequences.

These discoveries shook me to the core. Girvan was a messmate, a perfectly competent submariner and a commissioned officer. How on earth had he managed to con the Admiralty Interview Board and beat the close character scrutiny of the Dartmouth training regime? And how had he managed to fool his colleagues for four years in the intimately close confines of a submarine? He had been strangely unpopular but no one could have suspected the truth. Now he had simply disappeared.

It was almost six months before he was heard of again. He was eventually caught when the British Consul in Marseilles was suspicious about a man claiming to be a Royal Navy officer out of money and needing cash urgently for an airfare to Malta to join his ship. As a Brit had been suspected of a spate of thefts from yachts in the local marina, the Consul called the police. David was then charged with theft and taken into custody. His defence in the French courts, followed by extradition to face court martial in Britain, proved to be an expensive process, all funded by his unfortunate parents. He was duly court-martialled and dishonourably discharged from the Navy.

Perhaps I've been lucky. Probably I'm naïve. But I can't imagine any of the young men who joined the Navy with me or with whom I have served since, being guilty of such crass dishonesty. The Navy would like its officers to be paragons of virtue, true knights in shining armour. So would I. But in a large macho organisation like the Navy, there are bound to be some of the less gentlemanly sort. There will be bullies who would trample over you in pursuit of their own career and others who are merely hard-nosed bastards devoid of human compassion. There are some who would cheat on their wives and even try to cheat with your wife, but surely none who would resort to petty thievery and fantasy.

The sad thing about Girvan was that he had proved himself to be a competent submarine officer and had been accepted and trusted by his peers. As a well paid bachelor, he had no need to fabricate such a tissue of fantasies or steal. In James Thurber's book, Walter Mitty dies bravely at the hands of a firing squad. I don't know what became of David but he must have ruined his parents, both spiritually and financially.

****

After the Gibraltar Perisher, *Osiris* had a three-day recreational visit to Tangier. On arrival, I did the usual negotiation on the jetty for a ship-shore telephone connection with the representative from the local telephone company. Unusually, he was a rather smooth Moroccan in tweed sports jacket and flannels accompanied by a wireman. He was clearly of managerial status and without a second thought, I invited him on board for a whisky before remembering that Muslims didn't drink alcohol. When I apologised for the gaffe, he laughed and said: 'Don't worry. I'm a westernised Muslim. I'd love a whisky.'

His name was Mohamed – no surprise there – and we hit it off immediately. I guessed he was the manager of the local telephone company and had come down to ensure that things went well for the foreign visitor. When I gave him a conducted tour of the boat, he was so grateful that he invited me back to his flat for dinner and introduced me to his wife, and then he retrieved a crate of beer from under the sofa. In return, I invited them back on board the following evening and told them to bring some friends.

The following evening they arrived with two friends, one a very attractive young lady, a star of the local ladies football team. The other was a smooth, highly educated man in what was clearly an expensive black djellaba; I assumed he was linked to the attractive young woman. As we sat round the Wardroom table in an atmosphere of international goodwill, the subject turned to politics. I loved that. It was my favourite subject. But this was not like the usual sort of discussion one has with Muslims that almost inevitably end up with anti-Israeli diatribes. This discussion was calm, reasoned and analytical.

After about an hour, the smooth gent excused himself and left. As far as I could see, he had nothing to do with the glamorous young lady footballer. Nor did he seem to be close to Mohamed and his wife. I suspected that he was an agent of the Moroccan Secret Service. If so, he took back to his HQ an earful of my views on British foreign policy - not necessarily those of Her Britannic Majesty's Government.

Some months later, my mother-in-law in Uddingston answered her phone. It was a call from Morocco seeking Lieutenant Thompson. 'Oh,' she said, 'you mean the Admiral.' The Moroccan caller hung up. They could only have discovered my mother-in-law's phone number by tapping our ship-to-shore telephone calls. Kate and the boys had come over to Tangier and she had phoned home from the boat. Mohamed had been tapping our phone lines.

A few months later, there was an attempted coup against King Hassan II of Morocco, led by the Moroccan Minister of Defence and a group of Air Force officers. The plotters had attempted to shoot down the king's civilian aircraft when he was returning from France but failed. The Minister was then summarily executed. The other lead plotter escaped to Gibraltar and requested political asylum but was handed back to the Moroccan authorities and also executed despite a 'no-death-penalty' agreement with the UK Government. Eleven other Air Force officers were imprisoned.

In retrospect, I presume the call to my mother-in-law was a Moroccan Intelligence trawl to find out who had been talking to the British ahead of the coup. I wonder now if the smooth operator in the black djellaba had indeed been a Moroccan Intelligence officer or an Air Force rebel trying to assess likely British sympathy for his proposed coup.

****

On the long passage back to Faslane, an exhausted pigeon landed on our fin. We were its only alternative to drowning and the crew took it under their wing, so to speak. We released it later off Land's End.

By remarkable coincidence, a *Sea Harrier* from the aircraft carrier *Illustrious* suffered a very similar experience at the same time in the same area. The pilot lost contact with the carrier and was running out of fuel. He was about to ditch into the sea when he sighted a small Spanish cargo ship and to the astonishment of its crew, landed the Vertical-Take-Off-and-Landing *Sea Harrier* on its deck. The aircraft and pilot then travelled several thousand miles to the ship's next port-of-call before they could be offloaded, just like our pigeon.

Pigeons are no strangers to submarines. During the First World War when radio communication was still in its infancy, British submarines in the North Sea used homing pigeons to carry enemy sighting reports back to base when beyond radio range. The system worked reasonably well but was not without its technical hitches.

DOVES OF WAR

'Up periscope,' the Captain roared
And up the look-stick flew.
A fifteen second all round look,
Then turned to face his crew.

'The target is within our sights.
We need to signal back
And let the Admiralty know
We're going to attack.
Signalman, write that down in code
And fetch our feathered ace,
Then tie my signal to his leg
And send him back to base.'

The submarine rose from out the sea
Like prehistoric beast.
Its conning tower was opened quick.
The pigeon was released.

Arooga – Arooga – Arooga,
The klaxon sounded thrice
And to the deep the boat returned,
Its crew as cool as ice.

'The target's coming at us,'
Roared the Captain with delight,
And hugged the slim brass periscope
To keep it in his sight.

'Good God!' he groaned with tortured look.
'This is a bloody farce!
Our pigeon's landed on my 'scope.
And all I see's its arse!'

'A pigeon's not a seabird, Sir,'
The First Lieutenant said.
'It cannot land upon the waves
It's simply used its head.'

And every time the 'scope went down,
The bird flew round in search.
And every time the' scope went up,
It landed on its perch.

'Good God,' the angry Captain cried.
'How am I to attack,
When every time I raise my 'scope,
That bloody pigeon's back?'

'Perhaps,' the First Lieutenant said,
'He's waiting for his mate.
She's nesting in the engine room
Inside a piston crate.'

'Well get her out!' the Captain roared.
'Prepare the hen for flight.
And get the pair both on their way
So I can have my sight!'

'I'm sorry, Sir,' a stoker said,
'The hen has just been fed.
There's no way now she'll take the air.
She thinks it's time for bed!'

'God give me strength!' the Captain sighed.
'To fly is what they're for!
Has no one told our pigeons that
This country is at war?'

'I think,' the First Lieutenant said,
'To press home your attack,
We'll have to surface one more time
And get our pigeon back.'

****

When my time in *Osiris* was up, John Holl organised a farewell run ashore to the Victoria Hotel in Rothesay, the boat being at a buoy in the bay. The evening followed normal submarine practice. I drank too much, performed my party trick of appearing to fall down stairs, cracked my head on the landing wall, and lay dazed on the floor gazing up at a mural of an eagle with its wings spread. Below it was written the motto: *He who hoots with the owl by night cannot soar with the eagle by day* – a good message for life.

Thus ended my time in diesel submarines.

Trials and Tribulations

'A railway station?'

LADY BRACKNELL: OSCAR WILDE

I was now due for a shore job and my priority was to avoid another house move for Kate. She had a life to lead as well. My duty was clear. I had to go to Whitehall and beg for a job at Faslane.

My Appointer, the man who would decide my future, worked in the historic Ripley Block, built in 1726 and oldest of all the Admiralty Buildings in London. Ironically, it stood next to the Whitehall Theatre, home of British farce. Conscious of the building's history, I picked my way gingerly across its cobbled courtyard just as Nelson had done two hundred years before me and for the same reason, to plead for a job. In the entrance hall stood a fireplace and two antique, black-leather, hooded chairs from which hall porters would have leapt up to take Nelson's coat.

One approached one's Appointer with great trepidation. He knew all about you. He had read your staff reports; you had not. He had viewed the ultra 'Staff-in-Confidence' recommendations on your promotion prospects, penned by anonymous staff officers in the inner sanctums of the Navy, and he knew exactly where you stood in comparison with your peer group. For fast-trackers, he would be building the balanced careers required for high office; the rest of us were merely pawns. When visiting one chess-illiterate Appointer, I was dismayed to see my name inscribed on a piece of Lego halfway down a column of Lego bricks, each one representing an officer in my promotion batch. Nelson had been honoured with a whole column; I had achieved only one plastic brick.

Appointers were handpicked for their charm, diplomacy and communication skills but I was never fooled. The wise man knew very well that his Appointer spoke with a forked tongue. When I asked mine for a job in Faslane and preferably in torpedo development, he almost fainted with joy – nobody asked for that. Torpedoes were way down market and Faslane was a remote, rain-sodden, some would say god-forsaken outpost that none but a demented Scot would want. 'Yes, yes, yes,' he cried in exultation, 'I have just the job for you.' He was beside himself with delight.

I knew immediately that he had a dead-end job to fill but took it without a murmur. I was just the sort of yes-man Appointers liked. I had solved his problem. I was to join the *Mk24* Torpedo Trials Unit (TTU) based in the Royal Naval Armament Depot Coulport, the civilian-managed, top-secret ammunition depot built to handle Polaris missiles and their nuclear warheads. I was heading out of the mainstream into career no-man's-land but I would be living at home and had a mission to pursue. After all, I was passionate in my belief that we needed an effective new torpedo. I could see no point in having nuclear-powered, toothless bulldogs. This apparently dead-end job could even be a blessing in disguise, an opportunity to shine my light in the submarine wilderness.

On the morning of 25th October 1971, I joined the caterpillar of morning traffic crawling up the eastern shore of the Gareloch towards Faslane but this time I drove on past the high, barbed-wire security fence, past the adjacent ship-breaking yard where ships from the German High Seas Fleet had been broken up, past the Faslane cemetery where the victims of the *K13* submarine disaster lay buried in the shadows of turf-covered oil tanks, and through the village of Garelochhead to the opposite side of the loch. A third of the way down the shore road, I turned right on to the single-track Peaton Hill road and drove up over the spine of the Roseneath Peninsula.

At the crest of the hill, I stopped to savour the view. Six hundred feet below me lay the deep, windswept waters of Loch Long, so steep-sided that aircraft carriers and fully laden supertankers could tie up virtually against its rocky foreshore. Across the loch, the Cowal Hills rose two-and-a-half thousand feet straight up from the water's edge. To the North lay the Arrochar Alps and the West Highlands. To the South lay the gateway to the world. I could see straight down the Clyde estuary, through the Cumbraes Gap and beyond Arran to the crags of Ailsa Craig. There lay the North Channel between Ulster and Scotland, the entrance to the Atlantic. Nine hundred miles to the South, in a straight line down through the Irish Sea, lay the coast of Spain. How blessed I felt to be living and working in such a landscape.

I have always counted my blessings. I have led a charmed life. Despite my warlike occupation, I have known only peace; the same could not be said for the people of Ulster. It was less than three years since my visits to the Province in *Otter* and *Andrew* and now a virtual civil war was raging. British troops, all twelve-and-a-half thousand of them, had been sent in originally to keep the peace between warring Protestant Unionists and Catholic Irish Republicans. Now they were the targets. The Ulster Parliament had been suspended, Direct Rule from Westminster had been imposed, internment had been introduced and three thousand Irish Nationalists had fled to the Irish Republic. The IRA had established No Go areas in the Republican parts of Belfast, Newry and Londonderry, which the British army had then to invade. A year later, the infamous Bloody Sunday happened when British troops policing a civil rights march in Londonderry opened fire on what they thought to be IRA gunmen. Fourteen Nationalists were shot dead and the ramifications would rumble on for the next forty years. Bloody Sunday was followed by Bloody Monday and Bloody Friday when the IRA killed and maimed hundreds of innocent civilians.

All this was happening only seventy-five miles from my viewpoint. On a clear day, Ulster was visible. My sister and her husband were living there, Ray, now a teacher in Londonderry High School, and Sam working in the Du Pont chemical plant at Limavaddy. They were very happy living and working there but soon decided to get out; this was not their fight. As Ray remarked: 'I can't believe that such nice people can do such dreadful things to each other.' *Homo sapiens* could put man on the moon and prevent world war through nuclear deterrence but cannot stop fighting in the streets.

At the bottom of Peaton Hill, I turned on to the road to the Armament Depot, a mysterious place even for me as a serving submarine officer. Its entrance was underwhelming, no more than wire-mesh gates topped with barbed wire and guarded by Ministry of Defence Police. Beyond the gates, all that could be seen was a small office block with a couple of industrial warehouses lurking behind. However, up on the boundary fence, watchtowers manned by armed police punctuated the skyline and a battalion of Alsatian guard dogs were on the prowl.

One of the warehouses, known by the stirring title of Building 41, was to be my base for the next two-and-a-half years, its main purpose being to handle great wooden packing cases containing Polaris missile spares. It was the complete antidote to Ripley Block. This was the architecture of the Cold War. Torpedo Trials Unit had been given a small suite of offices on the mezzanine floor.

I have observed in life that there are sad buildings and happy buildings. Whitehall's majestic Ministry of Defence Main Building, carved in white Portland stone, was, in my experience, a sad building. Building 41 at Coulport with its breezeblock walls and corrugated iron roof was a happy building. In it, I was about to meet one of my greatest lifetime role models.

Commander Norman Jones, RN, was wearing a white lab coat over his uniform when I called to introduce myself. He was of slim build and medium height with a soft complexion, light brown curly hair, twinkling eyes, and the most engaging smile of any man I have ever met. He was softly spoken and utterly charming. Professionally, he was of my own breed but was about fifteen years my senior. He had qualified in the old Keyham College, Manadon's predecessor, and then served in *T-Class* submarines before moving into frigates in the Persian Gulf. He knew everyone in the building by name – cleaners, coffee ladies, secretaries, forklift truck drivers, civil servants, naval personnel and contractors – and he addressed them all by their first names as if they were his family. To command such a mixed bag of people, all with different agendas and vested interests, called for an entirely different approach to military leadership and Norman Jones had it taped. He was, in my eyes, the perfect officer and living proof that leading through courtesy and respect can be just as effective as ruling by diktat.

But leadership is not only downward looking. Norman faced an uphill challenge. His boss, the head of the Torpedo Project Executive in Bath, was a political animal. He had to be. He had to answer not only for the technical success of the project but also for the success of the political decision to privatise such a major Defence programme. *Tigerfish*, as the *Mk 24* torpedo was now called, was a political hot potato. Here was the old political trick for salvaging failure: change the management and reintroduce the project under a new name. Norman could easily be made the fall guy if our trials went awry.

As part of my joining routine, I called on the Director of Torpedo Design at the Admiralty Underwater Weapons Establishment at Portland, infamous for its spy scandals. He was a wily Admiralty Civil Servant whose department had designed *Tigerfish* in the first place but had been relieved of its project management responsibilities by the new Torpedo Project Executive after the weapon failed its Acceptance trials. He held Naval Officers in contempt. When I knelt before him, he explained to me that Naval Officers, like Julius Caesar, came, made a fuss, then left, whereas Civil Servants just kept rolling along. He saw us as ephemeral. He knew that troublesome officers would blow over after their two-year appointments. 'Troublesome,' to him, often meant reporting uncomfortable truths like trials not going well.

Another Jones, Lieutenant Commander Conrad Jones, was my immediate boss. He was responsible for the conduct of sea trials, a far more difficult task than the layman could ever imagine. Conrad had a pointed, piebald beard, curling moustache, and wore his cap at a jaunty angle like Admiral Beatty, hero of Jutland, giving the impression that he had just been walking the decks of a First World War battleship. He could easily have been mistaken for King George V or his murdered Victorian cousin, the Tsar of Russia.

Like his namesake, Conrad was a joy to work for and, despite his regal image, was modest and amiable with a deceptively diffident manner. At times he seemed to suffer from a lack of confidence over what was about to happen next, which led to the endearing habit of pausing mid-sentence as if he had just run out of certainties, then picking up the lost sentence with the phrase, 'the point being,' thus buying himself time to think of what his point could possibly be. This was hardly surprising as there was little scope for confidence in any aspect of our task. Most deadlines were missed, the torpedo's performance was erratic and we were under constant pressure to speed things up. Frequently, his meticulously planned trials had to be changed at the last minute or cancelled altogether.

Frustration seemed to be the order of the day and the patience of participating submarine Commanding Officers was often tested to the limit. On our first trial firing against a real submarine target, delay after delay prompted the following exchange between attacker and target.

Target to attacker: Isiah 6.11 ('How long, O Lord?')
Attacker to target: Ecclesiastes 5.2 ('God is in heaven and you are on earth, so let thy words be few.')

One frustrated submarine Commanding Officer even referred to our trials armada as *Monty Python's Floating Circus* after the anarchic television comedy series. Conrad thus earned the title of Monty Python and the name stuck. He took the joke with good grace but was not impressed when accosted by a woman in the Yacht Club in Gibraltar after deep-water trials.

'Are you with the circus?' she asked.

'Er…Yes,' Conrad replied.

Pause.

'You must be Monty Python!' she exclaimed in triumph.

As if the impatience of some submarine COs were not enough, Conrad was also constantly under pressure from the dynamic but abrasive project leader from Marconi Space Defence Systems. He had previously worked on an anti-aircraft missile project and simply refused to believe that sea trials with a torpedo were more difficult. At times he was completely unreasonable but eventually accepted that the weather was beyond Conrad's control.

My role was to be Conrad's winger on board the firing submarines. My technical partner in the circus ring was Barrie Sadler, a fellow Weapons Engineer with whom I shared an office. He was a great, noisy, bear of a man with flaming red hair and the war cry: 'You scratch my back and I'll ram the bat up your arse.' I never did summon the courage to tell him that it was a mixed metaphor. His job was to get round the trials armada and ensure that our miscellany of tracking equipment was fully operational.

As if to embellish our circus reputation, the torpedo handling party consisted of six sailors who went by the names of Black, White, Grey, Brown, Green and Ostrowski, the latter being the son of a Free Polish wartime hero. Ostrowski, we assumed, was Polish for purple.

The final member of the trials team was Jack Holmes, a venerable retired Lieutenant Commander who had joined the Navy as a Boy Seaman before I was born. Following his retirement from active service, he had been employed at the torpedo factory in nearby Alexandria. In deference to his longevity, I referred to radar by its pre-war title of RDF (Radio Direction Finding equipment). Jack provided an endless source of salty sea dits (Naval anecdotes) to entertain us during the long waiting periods.

****

The *Tigerfish* torpedo was designed purely for underwater warfare and was to be *the* weapon for a submarine dogfight. It was also a weapon of stealth, the dagger under our cloak. The first the enemy would know about *Tigerfish* was when it blew them apart. It was quite simply the most advanced torpedo in the world, so much so that its original code name had been *Ongar*, which had been the last station on the Central Line until London Transport closed it.

The problem with *Tigerfish* was that it didn't work. It suffered from so many technical hitches that it had failed Acceptance trials and had been on the brink of cancellation which would have destroyed the entire British torpedo industry. However, a Government report had concluded that it should be given a second chance with development being transferred to the private sector. Design had therefore been contracted out to Marconi with the mass production contract going to Plessey. We were to conduct the repeat Acceptance trials.

Tigerfish was a thing of beauty, if you liked torpedoes. It had a shiny, black, anodised aluminium body with a highly polished stainless steel nose adorned with a geometrically perfect array of listening devices and a streamlined tail with a beautiful pair of golden contra-rotating propellers. It was a heavyweight beast weighing in at one-and-a-half tons with a twenty-one inch waistline and twenty-one foot length (6.4 m). It was also wire-guided but unlike its predecessor, was prodigiously good at homing and had a massive twenty-mile (32 km) range.

Alas, this baby had serious teething problems, one being that its umbilical cord frequently severed at launch leaving it free to do its own thing – which it did. A typical problem was that when it surfaced at the end of its run, its ON/OFF switch would stick at ON allowing it to sprint round like a demented dolphin until its battery was exhausted, or it hit something. The Trials Unit had just been expelled from the Clyde for a number of embarrassing misdemeanours on the part of the weapon.

One rogue weapon had set off in pursuit of the Arran ferry. Another celebrated freedom by heading for the nearest beach, its sudden emergence from the sea so alarming golfers on the local links that the assault was reported in the national press. Fortunately, trial weapons carried data recorders and not warheads. This incident re-surfaced some years later in the hit BBC comedy series, *Yes Prime Minister*, in which Sir Humphrey, the loquaciously obfuscating Civil Servant, finishes a diatribe by adding, '...like the torpedo on the golf course, the less said the better.'

With its umbilical cord broken, *Tigerfish* was like a dog-of-war unleashed. It would attack the first thing it heard and one never knew what that would be. It was also perfectly capable of turning back and attacking its master, a suicidal flaw for a killer weapon. Just before expulsion from the Clyde, one *Tigerfish* did exactly that. It went sniffing round in a large circle, found nothing worth attacking, and returned to spear its parent submarine through the duck's arse, as the stern of a diesel submarine was called. *Ocelot* thus found herself with a jammed rudder being towed backwards into Faslane for repairs.

As taming this monster in the busy waters of the Clyde was far too risky, we were banished to the remote Sound of Raasay.

****

Not even the prodigious imagination of Sir Compton Mackenzie, author of *Whisky Galore*, could have conceived what happened next. Whilst space probes were sending photographs back from Jupiter and Mars, Americans were driving across the moon in lunar buggies and the Soviets were operating an orbital space station, the headquarters of Britain's latest top-secret torpedo project was being installed in the Waiting Room in Kyle of Lochalsh railway station on the fish pier, there being no other office accommodation available in the village. Rail passengers arriving at Kyle would now be confronted by an array of signs that read: *Toilets, Public Bar, Torpedo Trials Unit* and *Ticket Office.*

As the railway to Kyle had been constructed largely for transporting fish to the markets of the South, we now had to rub noses with railway passengers whilst our trials vessels had to compete with fishing boats for space on the pier. At weekends, the latter could be tied up five or six deep. Our spent torpedoes had thus to be landed along with fish catches in full view of the general public – and potentially Russian spies. NASA had it easy compared to us. It had a vast launch pad all to itself and simply fired rockets into space.

Predictably, in a tiny community like Kyle, the arrival of the Navy raised questions and there were immediate calls for a public meeting. This was duly held in the village hall, and the Commodore Clyde accompanied by Conrad took the platform to explain the Navy's need to test torpedoes in the deep, sheltered water of the Sound. The first question could have been foreseen.

An elderly fisherman speaking perfect English with a West Highland accent, stood up and asked: 'Does this mean the Navy can only fight in sheltered water?'

He was assured that the weapons being used were practice torpedoes and required sheltered water for subsequent recovery.

The Commodore had neither been briefed on the opposition, the likely questions, nor on how to deal with the local media. He was a Naval Officer, not a politician. Ranged against him were landowners, fishermen, the Free Church of Scotland, holiday-home owners, and the *West Highland Free Press,* all with axes to grind. Quietly in favour were the local bed-and-breakfast establishments, a couple of small hotels, and a handful of publicans, all of whom could sense business opportunities.

The landowners' objections were that Naval helicopters would disturb the rutting of their deer. The local fishermen were worried about the loss of their fishing grounds. The Free Church of Scotland was up in arms at the prospect of people working on the Sabbath Day – it was already opposed to the Skye ferry operating on Sundays.

Foremost amongst the interested parties was the *West Highland Free Press,* a unique local newspaper which had just been established by a young professional journalist called Brian Wilson who had already played a leading role in the protest against the basing of American Polaris submarines in the Holy Loch. He was a graduate of Dundee University and native of Dunoon, which neighboured the American base. Wilson had recently moved to Skye to set up his employee-owned newspaper with university friends. I had read it and was mightily impressed – it read more like a Highland version of *The Guardian* than a local rag. It was radical and political, with subjects like land ownership, Highland and Island economics, and the preservation of the Gaelic language featuring prominently.

None of us could have guessed at the calibre of these 'local' journalists. Wilson actually went on to write for *The Guardian* and other top national newspapers and became a Labour Member of Parliament and Government Minister. As Minister for Education and Industry in the Scottish Office, he would deliver many of the things for which the *West Highland Free Press* had campaigned. He would also go on, somewhat surprisingly, to campaign on behalf of the civil nuclear power industry, Hunterston nuclear power station being in his parliamentary constituency. He was a million miles away from being a wee parochial Scot.

I doubt if the Commodore or anyone else in the Navy had any idea of what we were up against. It was assumed, I fear, that the locals in such a remote place would have had only limited education and little knowledge of the wider world but I knew the score. I was a Scot and my highly educated father-in-law was a first language Gaelic speaker from a croft on Lewis. I knew very well that the locals of Kyle would be both canny and well educated.

At one stage during the meeting, when frustrated by the public interrogation, Conrad exclaimed in exasperation: 'You should all count yourselves damned lucky. If you were in the Soviet Union, you wouldn't be allowed to ask these questions!' He was right but it was not the recommended way of addressing a public meeting with the media in attendance. The following morning, a major Scottish tabloid bore the headline: '*DAMNED LUCKY WE'RE NOT SOVIETS' says Navy spokesman.*

Kyle of Lochalsh was then in the old Scottish county of Ross and Cromarty whereas Skye, on the southern side of the Sound of Raasay, lay in the county of Inverness. Triggered by the Kyle meeting, the residents of Skye also demanded a public meeting and this was duly held in the village hall at Portree, the island's capital. On the platform this time was the Member of Parliament for Inverness, a Skye County Councillor, and a Naval Captain, resplendent in uniform with four gold stripes on each arm. I attended with Mrs Branson, my landlady.

The hall was filled with about two hundred good souls who had braved the elements to come in from their outlying crofts and villages and the atmosphere was one of reverence, more like a Presbyterian Church service than a political meeting. There was little more than the occasional whisper in the congregation. When questions were asked, it was with quiet, respectful voices. Heckling was simply beyond the comprehension of such well-mannered people. The team from the *West Highland Free Press* were once again ensconced in the front row and asked most of the questions.

On this occasion, the Captain, sent up from the Admiralty Underwater Weapons Establishment (AUWE), knew what to expect. At least he did until an elderly, white-haired lady wrapped in a tartan plaid stood up and addressed him in a lilting

West Highland accent. She was almost certainly a first language Gaelic speaker and seemed almost to be crooning her question.

'I am Mrs Mackinnon,' she said with an air of polite confidentiality, 'and I have a croft on the island of Rona. What I am asking myself is this. Some years ago, the Navy put great big concrete obelisks on the shore at the bottom of my croft but we never had a public meeting about that. So what I am asking is: should I be worried by them?'

'Not at all, Mrs Mackinnon, not at all,' replied the Captain with great gusto. 'The purpose of the obelisks is, I'm afraid, a State secret, but I can assure you that they are absolutely nothing for you to worry about. The only people who need to be worried by them are the Russians.'

Mrs Mackinnon stood up again. 'Well, Captain, what I am asking myself now is that if they are secret, why would the Russians be worried by them when they don't know they're there?'

****

After the public meetings, the powers-that-be reverted to the Navy's standard method of winning friends and influencing people; we held a cocktail party. We invited the great and the good of the local area on board the RMAS *Whitehead* for the occasion (*Whitehead* was named after the inventor of the British torpedo who went on to work for the Austrian Emperor. His granddaughter married Captain Von Trapp, the World War 1 Austrian submarine ace subsequently immortalised in the Rodgers & Hammerstein musical *The Sound of Music*).

Whitehead was a three thousand ton research vessel owned by AUWE. It had a bow torpedo tube, deployable telescopic legs with listening hydrophones for torpedo tracking, workshops, laboratories and, crucially, merchant-navy-quality domestic facilities. The problem was that she was a little too big for the fish pier.

On the night of the cocktail party, *Whitehead*'s Captain decided that the weather was too wild to risk coming alongside and so anchored in the more sheltered waters of Balmacara Bay, some four miles away. This delayed the start of the cocktail party while Conrad resolved the logistics of getting guests out to the ship by boat in gale force winds and driving rain. In the interim, the guests were invited to refresh themselves in the public bar of the station at the Navy's expense, not the sort of watering hole the 'great and the good' of the area would normally frequent. Our Torpedo Handling Party, in best Number One sailor suits, acted as ushers but when the guests had all arrived, also took shelter in the station bar.

Having no responsibility for socialising, Leading Seaman Batty, a newly joined member of the Handling Party, reverted to type. He was a burly, bearded, beer-gutted matelot of some twenty years service with three Good Conduct chevrons stitched to his left sleeve and a silver Long Service and Good Conduct Medal dangling from his left breast. To the trained Naval eye, the medal coupled with his modest rank, indicated that he had long since reached his career peak.

Batty was drunk, not that anyone had noticed. Old soaks like him show little evidence of being under the influence and rarely cause trouble. He was a quiet, harmless type. I happened to be standing next to him making polite conversation with some of the guests when a loud shriek brought instant silence to the assembly. A local grand dame with a bedraggled ostrich feather in her turban had screamed in horror, thrown an arm across her eyes, and turned to face the wall. She had taken it upon herself to inspect Batty's uniform and he had responded by dropping his bell-bottoms to his ankles. He now stood, beer glass in hand, in full sailor suit to the waist but with only boxer shorts from the waist down, his well-pressed bell-bottoms now lying flat around his feet like a collapsed accordion. Despite his corpulence, Batty had ridiculously spindly legs and reminded me of Bluebottle in the legendary *Goon Show*. I confess, I knew not how to respond. The Navy, despite its centuries of corporate wisdom, had still not produced the drill for an officer to pull up a sailor's trousers.

Fortunately, the civilian skipper of *Thomas Grant*, a geriatric range vessel, had more presence of mind. He hoisted Batty's bell-bottoms back up to the waist and led him quietly behind the bar, a good move but an unfortunate choice of destination. Later that night, poor Batty was found seriously injured on the deck of a fishing boat. In the driving rain, he had staggered off the end of the pier.

****

A surprise awaited us on return to Coulport. The large analysis room at the end of our corridor had been occupied by six young Wrens who were deployed on various activities: one knitting, one reading a book on Advanced Algebra, the others poring over magazines or chatting. These were our Weapons Analysts. They were the crème de la crème of the Women's Royal Naval Service (WRNS) and all had A-level maths. They would be feeding the data from the tracking range and torpedo recorders into our analysis computer.

It was the first time since school that I had worked with females. Instinctively, I presumed them all to be innocent young ladies. Of course, I was wrong. One of them seemed to have hollow legs for she could drink any man under the table. Another was reputedly a man-eater and frequently arrived in the morning looking like she had been dragged through a hedge backwards. She must have been like a fox in a hen house living amongst the hundreds of sex-starved young submariners marooned at Faslane. As far as I could see, Wrens were simply female versions of sailors. They behaved like matelots, spoke naval slang, and shared the same Naval discipline.

The one with the cherubic smile reading the book on algebra was almost Home Counties stock. Like Conrad, she was a keen equestrian and was studying for promotion to officer. On her desk was a placard with the Roman motto: *Hic Fute Ludus Militorum*. As I had studied Latin at school, I thought I would impress her with my erudition. '*Hic* means *this*,' I said with a knowing smile. '*Ludus* is a *game* and *militorum* means *of the military*. So,' I concluded, '*hic ludus militorum* means *this game of the military*, but I can't remember the meaning of *fute*.'

She put down her book, fixed me with her dark-brown, doe-like eyes, smiled sweetly and said: 'Fuck this for a game of soldiers.'

In charge of the Wrens was Third Officer Pip Duncan, magnificent and irrepressible. She was of similar age to me and had worked her way up through the ranks from being a Weapons Analyst. She could be wild, witty and robust, and towered over me. She dealt with male officers with consummate ease and I doubt if any of her Wrens would ever have dared to cross her. Sensing that I was very correct and a tad shy, she took savage delight in planting herself on my lap in full view of the boss, knowing that it would embarrass me. It was pure devilment rather than anything sexual. Had a male officer sat on my knee, he would have been knuckled or court martialled. For a civilian lady to do it was unthinkable. When Pip did it, it was just for a laugh. That was the unique rapport that existed between male and female colleagues in the Royal Navy in the days of the Wrens.

The WRNS was created during the First World War but had been disbanded at the end of hostilities. It was resurrected in the Second World War and survived until 1993 when women were finally allowed to join the Navy on an equal footing with men. Yet they were still not allowed to serve in submarines and I could understand why. One submarine Captain was leaning over his chart table in dim red lighting when he felt a hand coming through between his legs and clutching his undercarriage. He spun round in disbelief to find himself staring straight into the eyes of an embarrassed-looking First Lieutenant who spluttered: 'I'm terribly sorry, Sir. I thought you were the Navigator!'

(When women were allowed to go to sea, a No Touching Rule was introduced.)

****

My lifestyle now can only be described as surrealistic. About twice a month I made the incredible journey from Coulport to Kyle of Lochalsh and back. This took me up the Bonnie Banks of Loch Lomond, over the wilds of Rannoch Moor, down through the brooding crags of Glencoe, up the Great Glen to Invergarry, over the high road to Glen Moriston, past the majestic Five Sisters of Kintail, and down to the iconic Eilean Donan castle before reaching the fabulous seascape at Lochalsh. Then it was over the sea to Skye on the old car ferry and into the welcoming home of George and Maureen Branson in Kyleakin with whom I stayed during trials.

As I was travelling throughout the year, I was able to witness Scotland in all its mantles. In winter, when driving back at night through avenues of ploughed snow, we would catch thousands of tiny red lights in the loom of our headlights. These were the eyes of unseen herds of red deer staring back at us; they had come down to the roadside in search of food. In Spring, the banks of Loch Lomond would be covered in a carpet of bluebells. Autumn, my favourite season, would begin with the hills glowing purple with heather and end with the glens turning gold and brown. Never mind torpedoes, this was living.

As we had to maximise daylight on task, trials routine was exhausting. We would leave Kyle by boat long before breakfast and board the submarine out in the Sound, a sixty-minute boat trip but in more breath-taking scenery. To the West, the jagged ridge of the Black Cuillins rose behind the low-lying island of Scalpay, to the East lay the wilds of Torridon, and to the South, over the Kyleakin peninsula, lay the high peaks of Kintail and Knoydart. The seascape was every bit as entrancing. The Sound was teeming with wild life. Seals were commonplace. Seabirds were everywhere. I even watched a pod of Killer Whales go past.

Descending into the warm fug of the firing submarine for breakfast felt like coming home but finding a seat in the tiny Wardroom was a perennial problem. We could be out for twelve hours with no bunk or office into which we could retreat. On return to Kyle, we would debrief in the station Waiting Room with the other players. I could be away from six in the morning till ten at night when Maureen Branson would still insist on feeding me. She loved having the Navy around. It gave her a little bit of extra income but she also just liked 'the Navy boys' as she called us. We were all very well behaved, caused no trouble, and were scattered round the village in lodgings; so the community took us to its heart.

****

On firing days, our small armada would assemble out in the Sound. It consisted of the firing submarine, occasionally a target submarine, an old Inshore Minesweeper converted for torpedo recovery, a Vickers-operated ship with a DSRV (Deep Submersible Remote Vehicle) for sunken torpedo recovery, *Whitehead* whose built-in torpedo tube was used for firings when no submarine was available, *Thomas Grant,* an antiquated torpedo carrier Conrad had unearthed in Portsmouth Dockyard and which Barry had converted for trials purposes, and three Fleet Tenders that acted as water taxis and guarded the approaches to the Sound during firings. This was our *Floating Circus*, a gathering of the maritime clans worthy of the Lord of the Isles.

There were so many things that could go wrong during trials that it was miraculous if a day's firings were completed according to programme. First, all the craft had to arrive on time, all being hostages to weather, defects and spare part availability. The weather was also critical for torpedo recovery and on the West Coast of Scotland, it seemed to be wilfully against us. Weather permitting, trials could not then commence until tracking devices had been electronically synchronised over dodgy radio links. Then the torpedo and submarine weapon system had to pass last minute readiness tests which frequently threw up problems. After that, a long countdown would begin, just like for a moon shot. Then Murphy's Law would strike and a Fleet Tender would report a ship approaching. That meant a thirty-minute delay while it transited the Sound, by which time the weather window would have gone. (Murphy's Law: *If things can go wrong, they will.*)

Inevitably, resumption of the countdown would occur just as we were about to have lunch. On one occasion, Lieutenant Phil Cregeen, the submarine's Weapons Officer, asked me to substitute for him while he had lunch, so I manned the firing panel. The Control Room was packed solid with bodies, the Captain holding centre stage on the periscope with Marconi's trials officer at his side. When I heard the word 'Fire' I pressed the fire button and the submarine shook as one-and-half tons of torpedo was launched.

'Who fired that?' the Captain roared in horror.

'I did.'

'I ordered do not fire!'

'I only heard 'fire.'"

In the Navy, the order not to fire is: *Check Check Check.* It is not: *Do not fire* because, if the man with his finger on the trigger hears only the word 'fire', he will. Torpedo recovery then became the critical factor. The next firing could not take place until the previous weapon had been recovered, which was an operation in its own right. At the end of its run, the torpedo was designed to drop ballast, become buoyant, and bob up to the surface, but its little flat head was very difficult to see if the sea was anything other than flat calm. So it emitted pings that could be detected on sonar, but often the pinging could not be heard. As one wag remarked when a weapon was lost: 'Pings ain't what they used to be.'

When the dead weapon had been located, a Gemini rigid inflatable would be deployed from which the sea-rider would carefully clamp a protective nose cage on to the torpedo's ultra-sensitive head to allow it to be hauled out of the water without causing any damage. Unlike a warshot, an exercise weapon had to be re-used and the slightest scratch on its nose would create flow noise next to its highly sensitive ears, thus causing it to go round in circles like a dog chasing its tail.

At the end of one very long day, Murphy's Law struck again. Our Torpedo Recovery Vessel, complete with a load of top-secret torpedoes, ran aground on an outlying spit of sand off Scalpay. In fact, Murphy struck thrice. Second, the tide was ebbing and the boat was left high-and-dry until the next high tide when it could be re-floated. Third, Scalpay was home to one of the principal objectors to our activities. A boat full of torpedoes at the bottom of his garden was just the metaphorical ammunition he needed. Armed with a camera, he walked out to the stricken vessel at low tide, photographed its cargo and promptly sent his pictures to the national press. The following day they were front-page news. But it was not all bad. The torpedoes still had nose cages fitted and that gave them a particularly sinister look, as if they had some sort of gaping mouths at their front ends like jet engines. Inadvertently, we had just sold Soviet Naval Intelligence a brilliant dummy.

****

No nuclear submarine had yet fired a *Tigerfish* and it was vital to prove they could. They differed from diesel boats in so many critical ways: bow profile, launch system, speed, etc. First *Warspite* and then *Conqueror* arrived and both conducted entirely successful firings. I then had to ride *Churchill*, our latest nuclear boat, from Coulport to Raasay to monitor a clutch of our eggs. It was my first experience of living on board a nuclear submarine and it was a different world.

Churchill was approximately the same length as a diesel boat but her teardrop hull gave her an extra seven feet (2.2 m) in diameter. That allowed two main decks and loads more space into which equipment could be crammed. She had a Commander in command and half her officers were Engineers, four being required to manage the nuclear propulsion plant, two of them being Lieutenant Commanders. In a nuclear boat, Engineers were no longer bit-players.

As the nuclear authorities demanded graduate engineers for reactor plant management, I knew three of *Churchill's* four Engineers. One of them, Rob Walmsley, was an exceptionally bright Cambridge graduate on a meteoric rise to the very top. He would go on to become Controller of the Navy and then Chief of Defence Procurement, the technical equivalent of Chief of the Defence Staff. The culture on board was also entirely different from a diesel boat. Nuclear boats had gravitas. The Engineer Officers wore white overalls and had blue radiographic film badges dangling like medals from their breast pockets to record their radiation exposure. They kept watch behind the mysterious Checkpoint Charlie, the reactor compartment bulkhead that lay about one third of the way along the hull and beyond which only nuclear-qualified personnel were allowed to go. *Churchill's* engineers seemed like atomic scientists going to work. Once or twice a watch, a message would ring out over the ship's public address system informing us that a nuclear procedure was taking place. It was all very awe-inspiring.

My task seemed very trivial in this atmosphere. I just had to measure the state of the torpedoes and confirm they were not being adversely affected by the pitch and roll of a teardrop hull on surface passage. As there were no spare bunks, I also had to sleep in an empty torpedo rack beside them, praying that I would not be launched whilst asleep.

****

Norman Jones had done a brilliant job in melding together his disparate team and harmony reigned in Building 41 but things were not so sweet on the commercial side. Industrial warfare had broken out. As Design Authority, Marconi was an integral part of our team but Plessey, the manufacturers, were over the horizon and had caused local political uproar by shutting down the Torpedo Factory in Alexandria and transferring all the work to their factory in Essex. Marconi though soon dispossessed Plessey of its lucrative production contract and achieved a monopoly of all *Tigerfish* contracts. But never forget, monopolies are not to be trusted.

Marconi's monopoly included reporting the results of our trials and they were not going to shoot themselves in the commercial foot by reporting failure. Some problems were thus kicked to touch with the argument that they would be solved in a successor *Tigerfish Mod 1* contract. This was all very satisfactory for Marconi's shareholders but we in the Navy needed a weapon that would work today rather than tomorrow. This led to frequent heated arguments over the interpretation of results. Our combined efforts, however, achieved nominal 'success' and *Tigerfish Mod 0* was duly accepted into service, warts and all. It had to be. Had it failed, the project would have been cancelled, the British torpedo industry would have died, and our submarines would have been left without an effective anti-submarine torpedo. We would then have been forced to buy an off-the-shelf American weapon, which our struggling national economy could ill afford. Besides military imperatives, there were political and industrial motivations too for *Tigerfish* to survive.

The country was in a mess. The trade unions were at war with the Conservative Government led by Ted Heath who had been forced to declare Emergency Powers to deal with a miners' strike that threatened to shut down national electricity supplies. This in turn led to a national three-day working week to conserve coal stocks at the power stations. A national dock strike followed. In February 1974, Heath resigned and lost the consequent General Election. Harold Wilson and Labour were then returned to power; the trade unions had won. Wilson then opted for a snap General Election and was returned with a larger majority for his third term as Prime Minister. Heath's defeat would have been music to Soviet ears. They may even have written the script given the fact that a number of key trade union leaders were Communists and Soviet sympathisers. In the meantime, we submariners would have to fight any war with the weapons we had.

The reality is that equipping the Armed Forces in peacetime is an exercise in pragmatism. The Defence budget is always tight, usually being squeezed, and large contracts come along only at widely spaced intervals – twenty years for a new torpedo – yet the Defence industries must be kept alive in the interim.

The cynic would say that the solution for the Defence industry is never to perfect a *Mod 0* product but make it just good enough to avoid cancellation, but still bad enough to require a follow-on *Mod 1* remedial contract. Acceptance trial reports thus become masterpieces of weasel wording and half-truths. The still imperfect *Mod 1* will then require further de-bugging but by this time too much has been invested and project cancellation would embarrass Ministers leading to a *Mod 2* development, and so on. Thus, employment is maintained, skills are retained, manufacturer's profits are returned, the Ministry is not embarrassed, and the weapon achieves perfection just in time to be scrapped and replaced by the first unreliable version of its successor. Military market forces apply only when we go to war. That is why at the beginning of most wars, personnel on the front line have to fight with imperfect equipment. It's then a game of catch up.

CHAPTER 12

Going Nuclear

E equals MC squared.
Our Albert worked that out.
It's such a lot of energy
To get from almost nowt.

AUTHOR

I was now uniquely well qualified in guided torpedo matters and proposed to my Appointer that I should become the founder member of a new specialist group of Underwater Warfare Engineers. I was ahead of my time however as it would be another ten years before such a group was formed. My Appointer wasn't keen on this idea anyway as he was under pressure to round up all the stray graduate engineers in the Submarine Service to man our fast expanding flotilla of nuclear submarines. I was to train for Her Majesty's Nuclear Service.

'It was,' he said, 'a case of all hands to the pumps.' I was to become a nuclear propulsion specialist, a creature of the engine room, not of the attack team. I was to be legs, not teeth. The brightest of my contemporaries had gone nuclear six years before and were about to become Senior Engineers of nuclear submarines. They were seasoned experts. I would now have to enter their field as a complete beginner.

Reluctantly, Kate and I packed our bags and moved from Faslane to Blackheath, conveniently close to the Naval College in Greenwich. It was an ideal location. Not only did it retain the sense of being a village within a great city but Blackheath Common also enjoyed the fresh air of Shooters Hill, one of the highest points in London. Charles Dickens referred to coaches lumbering up it en route for Dover in his *Tale of Two Cities*. The Common was also contiguous with Greenwich Park at the top of which stood the Royal Observatory, founded in 1675. Nestling below that at river level stood Queen Anne's mansion, built in 1616 and now the National Maritime Museum. Between the Museum and the Thames lay Christopher Wren's magnificent Naval College, built originally as the Royal Hospital for Seamen in 1696 but converted to a Naval College in 1873. The sense of maritime history in Greenwich was tangible.

From Blackheath to my classroom took only ten minutes by bike without having to run the gauntlet of London traffic. It was virtually straight down the Greenwich meridian and depending on which path I took, I could travel in either the eastern or western hemisphere. The great irony was that I was cycling through the 'nuclear free' Royal Borough of Greenwich, knowing that in the bowels of King William block in Wren's masterpiece sat *Jason*, a small nuclear reactor used by the Navy for training purposes. It was a sort of test tube reactor, used to illustrate the technique of 'going critical' i.e. sustaining a controlled chain reaction. (Enrico Fermi's first ever man-made chain reaction, achieved in the year of my birth, had taken place beneath a college football field in Chicago.)

The Naval College was a breath-taking place in which to study and its motto – *Tam minerva quam marte (By wisdom as much as war)* – was most apt for the Age of Strategic Nuclear Deterrence. One of its four massive blocks housed the Nuclear Department with a civilian professor in charge, another was home to the Second Sea Lord (Head of Naval Personnel), a third housed the Royal Naval Staff College and the WRNS Officer Training School, while the fourth was an accommodation block. Often, as I wandered out of some brain-stretching lecture on atomic theory, a squad of glamorous young Wren Officers would come marching past. All of us dined in the famous Painted Hall, its fabulous ceiling reflecting the triumph of Protestantism over Catholicism. Compared with this masterpiece of baroque sectarian art, the Orange lodges in the West of Scotland seemed very primitive indeed.

At Manadon, I had found Engineering utterly boring. To my surprise, I found the Nuclear Course stimulating. It was Applied Science at its best in that it brought together virtually all of the engineering disciplines. To understand the origins of nuclear energy, we were first taught Atomic Physics. That was fascinating. I came to understand Einstein's famous $E=mc^2$ formula. In simple terms, it means that if one splits the nucleus of an atom and then weighs all the broken bits, they will weigh less than the original, hence the saying: *The whole is greater than the sum of the parts.* The missing mass is the energy; the E in Einstein's equation is nuclear power. The trick is to continuously split atoms in a controlled chain reaction and harvest that energy without creating a bomb. It is exceptionally clever.

Having thus harnessed the nuclear energy in the form of heat, one has then to extract it immediately or the whole ensemble will 'melt down', a phrase that reactor technology has given to the English language. All the dry subjects studied at Manadon now came into their own: Metallurgy provided the special alloys required to create leak-free fuel cells containing the highly radioactive fission products, Thermodynamics and Fluid Mechanics saw to the transference of heat before the fuel cells melted, and by using the steam produced to drive turbines, Mechanics delivered useful work and the electricity thus generated enabled Electrical Engineering to provide reactor control systems and keep the lights on. Stick all this inside the hull of a submarine and one has the most amazing creation, a nuclear-powered fish. Even as a reluctant engineer, I found this inspiring.

The new dimension was Health Physics, the science of the effects of radiation on the human body. That was more sobering. Radiation induces cancer and severe doses are fatal. I learned about the different types of radiation and how to deal with them, about the statistical body tolerance for doses of radiation and about British Standard human organs. These Standard sizes were used in research. Some radioactive particles, for example, congregate naturally in the thyroid gland, so the British Standard Thyroid is the reference size for experimental calculations. Other radioactive particles are bone-seekers. It was all a bit worrying and left me in no doubt that radiation had to be treated with the greatest of respect.

Being of a lavatorial mind, I was amused to find that there was a British Standard stool. My curiosity had to be satisfied. Just how big was it? By my calculation, it was half-a-metre long! I have felt inadequate ever since.

Submariners instinctively take great interest in matters of the bowel. An Engineer Officer in *Resolution*, our first Polaris submarine, found himself a guinea pig in one of the many biological trials being conducted on our early nuclear submariners. The trial required every member of the crew to submit a stool sample before and after a two-month patrol. Kaiser Bill, as the officer was known, thought it would be amusing to submit a sample from his pet Labrador for his pre-patrol offering. That triggered a flutter in the research dovecots. Medical analysis concluded that one of *Resolution*'s crew was of non-human origin.

The College may have been a haven of nuclear academia wrapped in a cloak of history, but history was not standing still. In the world outside, France had recommended nuclear testing in the Pacific and India had just joined the Nuclear Club by detonating its first atomic bomb under the splendidly euphemistic title of *Smiling Buddha*. Now there were six in the Club. As a counter to India's new capability, Pakistan was also forging ahead with its nuclear weapon project. Few saw India as a threat to world peace but Pakistan was a different proposition. It was one of the world's most populous Muslim countries and was politically unstable. The prospect of an Islamic Bomb was a nightmare scenario, considering that Islamic extremists show no respect for human life and are vehemently in favour of destroying Israel. Small wonder then that Israel ensured that it too had nuclear weapons.

Israel had just been attacked by an alliance of Egypt and Syria in the fourth Arab-Israeli war, the Yom Kippur war. It had won that war emphatically and probably always would when defending its own turf, but an Islamic nuclear bomb dropping out of the sky on Tel Aviv was a different argument. Israel is surrounded by Islamic nations so inter-continental ballistic missiles would not be required, nor even intermediate-range ones. A short-range missile would suffice. It was little wonder then that Israel was about to become the seventh member of the Nuclear Club.

Israel can never afford to drop its nuclear guard. For the Israelis, it is about national survival. In Britain, as the memory of World War recedes from human memory, the need for self-defence becomes increasingly difficult to sell. In 1974, the British public was far more concerned with sugar and toilet roll famines.

The sugar crisis was triggered by a combination of Caribbean exporters switching to more lucrative American markets as the pound sterling fell in value, plus restrictions in home sugar beet production imposed by the European Economic Union, which Britain had just joined. A minor reduction in personal sugar consumption would have averted the crisis but that is not the nature of the human beast. Panic buying began. Fighting for sugar took place in supermarket queues. In Coatbridge, Mrs Lindsay, my mother's elderly neighbour, had her ankle broken in a sugar queue. Shops were cleaned out. There was no sugar left in Britain. The wafer thin veneer of civilisation had been ripped away. This was exactly what would happen after nuclear Armageddon.

The Great Toilet Roll Famine followed hot on the heels of the sugar crisis. I've no idea what started it – perhaps an epidemic of diarrhoea somewhere else in the global marketplace – but panic buying also cleared the shops of bog rolls. My neighbour in Blackheath, a well-educated Scottish dentist, took me into his garage and proudly showed me his personal horde. He had carton upon carton of toilet rolls stacked up to his garage roof, enough to wipe his bottom for the rest of his life. Why can't people be reasonable? Because, in a crisis, mob mentality rules.

My personal crisis was Kate discovering a lump in her right breast. The following morning, I left for the College with dire thoughts while she went to consult a local GP. When I returned home that evening, she was not there. She had left a note. She had been admitted to Lewisham General as an emergency case. My whole world was toppling. I rushed to the hospital and found her sitting up in bed in a large ward talking cheerfully to an Afro-Caribbean lady in the neighbouring bed.

'I'll arrange for you to go private,' I said. Despite our limited funds, I had enrolled her in BUPA, a private health insurance scheme.

'No need,' she replied. 'I've already had the operation. They've removed the lump.'

The National Health Service had beaten me to the draw.

'I'll arrange a private room then.'

'No need, I'm happy here. I prefer company.'

That was Kate, ever cheerful in adversity, and friendly with everyone.

The experience caused me to reflect on the virtue of our public health system. The husband of the Afro-Caribbean lady would have been suffering the same anxieties as I. What an uncivilised country it would be if our loved ones could not have medical care because it was unaffordable.

The lump proved to be no more than a blocked milk-duct and Kate was given the all clear but the incident left me in no doubt about my priorities in life.

The Nuclear Course was hard work. Towards the end with final exams looming, my tutor sent for me: 'Do you think you're going to pip it?' he asked. Clearly his money was on my failure.

Piss off, I thought. Why am I never credited with the will to succeed? 'Yes,' I replied. 'I told my Appointer this was beyond me. It was his mistake to send me here. If I fail, I prove my point.'

If I did fail, it would not have been for want of trying. I was working my butt off at home trying to get my brain round the things I had failed to grasp in class. This was usually because I was half-a-blackboard behind the lecturer, desperately scribbling down all the hieroglyphics before they were erased, my principle being to capture the lifeline of the lecturer's notes and study them in private later. Though I never managed to keep up with the hunt, my method worked and I passed comfortably.

Richard, my elder son, was less fortunate. He had to attend a local primary school. Every afternoon he came home in tears. It was a disaster. I feared bullying and went to see his teacher. She explained that in his kindergarten in Helensburgh, he had been taught to read in the old-fashioned way. In Blackheath, they were teaching by the modern phonetic way, such as 'school' being 'scwl'. It would have fooled me. The teacher explained that with a class of over thirty and Richard being the only odd-one-out, she could not give him the personal attention he needed, and so my poor five-year-old son had become a little lost soul. That was another lesson at Greenwich – children have needs that are not linked to their parents' careers.

****

For the practical part of my nuclear training, I was despatched to Dounreay on the North Coast of Scotland. At least I was. Kate and the boys returned to Helensburgh.

The great steel containment sphere of the UK Atomic Energy Authority research reactor at Dounreay was a national icon but few were aware that alongside it, the Royal Navy also had a research establishment, HMS *Vulcan*, operated by Rolls Royce and Associates. It contained a complete submarine nuclear propulsion plant with a simulator for operator training. In that simulator, we would be put through our paces. For me, it was a torture chamber – I was terrified of getting it wrong. All my old *Otter* anxieties came flooding back. Confidence is such a critical factor: with it, one can fly; without it, one is crippled.

Simulator training is necessary because it is not possible to rehearse nuclear emergencies on a real plant. The simulator was an exact replica of a submarine Manoeuvring Room and behaved like the real thing. Nuclear-qualified instructors hidden behind glass screens controlled it whilst trainees filled the watch-keeping positions. The instructors' role was to throw every imaginable plant emergency at us and just as the trainee Officer of the Watch was getting on top of the situation, to throw in another googly. The instructors' task was to put the trainee under extreme pressure to prove that he could keep a cool head in any emergency and deal with it while keeping the submarine operational. The simulator had everything a psychological sadist could desire along with all the ingredients for breaking a man. No matter what action the trainee took, the instructors could make it fail. It is amazing how the imagination takes over during simulation. Everything becomes reality, the heart pounds, and the adrenaline flows. It induced in me the terror of making a mistake.

As with the nuclear school in Greenwich, the location of the Dounreay nuclear plants on the shores of the Pentland Firth, surrounded by the desolate moorlands of Caithness, seemed utterly incongruous. This was Viking country. Thurso, the nearest town, is named after the Norse god Thor. On the moors behind it stood Pictish brochs, the circular Stone Age houses built for defence against Viking raiders. Remote the area may have been but it was even more historic than Greenwich.

I was accustomed to the West, South and East coasts where one senses land over the horizon. It felt strange to be looking to the North knowing that there was nothing but a few scattered islands between Caithness and the Arctic ice cap. One day I jogged over the moors and stood beside a broch, gazing out across the Pentland Firth and trying to imagine the panic in the tribe on sighting a fleet of Viking longships approaching.

Thurso, the most northerly town in mainland Britain, had a population of eight thousand people. Originally a fishing and farming community, it had gained massive importance during the two World Wars because Scapa Flow, across the Firth in the Orkneys, was a main base for the British Fleet. Thurso had been the arrival and departure point for hundreds of thousands of British sailors. The Navy no longer used Scapa Flow but the nuclear research establishments had brought new prosperity. On my first excursion into the town, I asked the local grocer if Thurso, being so remote, had been hit hard by the sugar crisis. He looked at me as if I were barmy. 'What sugar crisis?' he asked. 'We had no sugar crisis here. I know how much everybody normally takes and that's what they got – and no more.'

Kate phoned. She was unhappy about something; I can't remember what. Our marriage positively effervesced with friendly argument; it is the healthiest way to be. The only couple to divorce amongst my contemporaries had always addressed each other as 'Darling' and never argued. After their divorce, his wife confided that she had never dared to disagree with her ex-husband but Kate was made of sterner stuff.

I felt moved to send her a bouquet of flowers by Interflora but whilst in the florist-cum-greengrocer's shop, remembered that the florist in Helensburgh was also the greengrocer. 'Is it possible to send vegetables?' I asked. The shop assistant looked surprised but after a quick phone call, confirmed that it was. So Kate received a bunch of rhubarb via Interflora. That's what I call: 'saying it with flowers.' It's the way that you say it after all.

It was at the end of the Dounreay course that I first encountered the Big Brother effect. This stemmed from Admiral Hyman J. Rickover, USN, father of the nuclear submarine. His influence in the nuclear submarine world was such that it had pervaded our own nuclear propulsion programme. Since its inception, Rickover had run the US Navy's nuclear programme with a rod of iron, interviewing personally every one of their nuclear candidates. His was a zero tolerance regime and he had insisted on it being adopted by the Royal Navy when it was licensed to buy American reactor designs. Everyone was in fear of getting it wrong, and causing the first nuclear accident. In practice, this led to a culture of 'clearing one's yardarm', a naval metaphor meaning that if things went wrong, it would be someone else's fault.

At the end of the course, the Officer-in-Charge sent for me and declared that he was considering failing me. 'Go ahead,' I said. 'I've done my best. If it's not good enough, so be it. I told my Appointer this was not for me.'

I had called his bluff. If he failed me, he would have to explain why his training regime had failed to qualify a competent officer. If he passed me, he would run the risk of being blamed for delivering an incompetent. He was hedging his bets by putting me down as a potential risk; so, if I were to foul up, he could claim to have warned me. I passed.

I had now been baptised into another new religion. In this, the nuclear reactor was God, Rickover was Christ, and the Devil was incompetence. I was now its humble penitent. It even had its own form of confessional. From now on, all faults, failures and abnormalities had to be declared in written Incident Reports that would be scrutinised by a Spanish-type Inquisition known as the Central Plant Control Authority. If making a mistake was a sin, failing to report it called for burning at the stake.

From Dounreay, I was appointed to *Conqueror* to complete my nuclear sea training.

CHAPTER 13

Things That Go Bump in the Night

'From ghoulies and ghosties and long-leggedy beasties,
And things that go bump in the night, Good Lord, deliver us!'

TRADITIONAL SCOTTISH POEM

Conqueror lay low in the water alongside the jetty at Faslane. She was like a giant whale, her bulk hidden beneath the surface. Diesel submarines had a more swash-buckling appearance – ship-like bows, a flat superficial casing like a deck, and a rectangular fin. A nuclear submarine looks altogether more sinister, not like a ship at all. In place of a bow, *Conqueror* had a great, round, bull nose and her teardrop hull was streamlined. Her fin looked like an aerofoil and astern, almost as if a separate structure, her giant rudder stood tall like a whale's tail, only vertical.

On the surface, it is the nuclear submarine's fin that slices through the waves. The Officer of the Watch is in a wet and dangerous place in that position, like an old-fashioned figurehead. A lookout in *Courageous* had recently been sucked out of her fin in extreme weather as the boat literally went through a giant wave. He was never found. Since then, lookouts and Officers of the Watch have been belted in.

My first impressions on going down through *Conqueror's* accommodation space hatch were the size and smell. She was so much larger than a diesel boat yet there seemed to be little more space, it being crammed full of equipment. The ambient aroma was no longer of diesel oil. It was more of stale cooking and the sterile air of forced ventilation. On board, I received the warmest of welcomes; they had been looking forward to my joining. The wives had even sent Kate a welcome card. I could tell immediately that *Conqueror* was a happy ship. The feeling comes, of course, from the people, not the metalwork, but one always credits a ship with a character. Inexplicably, it seems to remain even when the people have changed.

There were five Engineer Officers on board, all ex-Manadon. Better still, 'Jumping Jim' Jones, the Electrical Officer, the role for which I was destined, had joined the Navy with me and we had always been chummy. He was extremely bright, ultra enthusiastic, bubbling over with knowledge and like the Light Infantry marching, talked at twice the normal pace. He was very much in the nuclear first eleven. Paul Anthony, the Senior Engineer, was a 'Clanky', a Mechanical Engineer, and was as taciturn as Jim was exuberant. He had recruited me into the cross-country running team at Manadon. I had also played in the College football team with John Board. I was very much among friends and none harboured any doubts about my ability.

The two younger Engineers, Chris McLement and Graham Stevens, were also extremely friendly. The fact that they were five years my junior but with greater nuclear experience underlined my predicament. If I qualified, I would end up being in charge of a whole team of engineers, all with more nuclear experience than I.

The Commanding Officer, Commander John Round-Turner, was, like Conrad, a keen equestrian. He seemed very jolly and made me feel instantly accepted. Lieutenant Commander Chris Meyer, the Executive Officer, had been in command of *Auriga* when she had her battery explosion and was a gem of a bloke. (In nuclear boats, the old title of First Lieutenant had been dropped in favour of Executive Officer). In his role as Mess President, Chris had mastered the art of ensuring that there was no sectarian divide in the Wardroom between the Engineers and Seamen who were like two separate clans. He was also the sort who shielded subordinate officers from the Captain's wrath. Tony Taylor, the Navigator, I knew from torpedo trials and Chris Lowe, the Sonar Officer, shared my offbeat sense of humour. This was a good team.

The social framework in a nuclear submarine Wardroom was utterly different from that in a diesel boat. For a start, the officers were very much more senior and there was no drinking culture. Professionally, it was as if two crews had been rolled into one. The Engineers, 'Back Afties' as they were known, lived largely in a world of their own, half the crew belonging to them. They operated under their own strict nuclear rules and the Senior Engineer was equivalent to the Executive Officer, except in the latter's capacity as Deputy Commanding Officer.

If the Senior Engineer considered something to be prejudicial to nuclear safety, he had the power of veto, the only exception being if the submarine itself was at risk. In principle, he could refuse to obey an order from the Captain if it threatened nuclear safety but in practice, that situation was unlikely to arise since the Seaman officers also undertook nuclear training and knew the score. Ultimately, the Captain was responsible for the safety of the ship, including nuclear safety, and was most unlikely to counter advice from his Senior Engineer. For his part, the Senior Engineer had to remain cognizant of the needs of the Command. As an ex-diesel submariner, I was thoroughly imbued with both cultures.

The Seaman Officers managed the warfare and domestic arrangements from the Control Room in the front end of the boat where the periscopes, sensors and weapon systems were. The Engineers operated the power plant from the Manoeuvring Room back aft. The arrangement reminded me of the gargantuan dinosaur Diplodocus, which was such an enormous creature that it required two brains. The first was in its head for organising feeding and navigation, and the second was above its hind legs to manage bowels and perambulation. The problem for Diplodocus was that its two brains were interconnected by a primitive reptilian nervous system that became sluggish in the cold and, like a government office, shut down completely after five o'clock. My theory for its extinction is that when the front brain spotted a predatory Tyrannosaurus Rex and the weather was cold, the order to run away didn't reach the after brain until it was too late. It seemed a good metaphor for the nuclear submarine where good communications between head and tail were essential for survival.

Jumping Jim, so called because of his irrepressible enthusiasm, took me under his wing immediately and gave me a tour of the boat, which was getting ready for sea. After we passed through the Control Room with its periscopes, chart table, and ship control systems, we came to the massive steel door of Checkpoint Charlie, the forward entrance to the Reactor Compartment tunnel. Jim announced: 'This is Grumpy Corner. It's where the on-watch wreckers sit. In there is the Elephant Snot Tank.' He pointed at a door.

As nuclear submarines spend months on dived patrols, the enclosed volume of air inside the boat has to be continuously re-circulated and refreshed. On its way round the circuit, it is passed through carbon dioxide scrubbers to save the crew from suffocation, oxygen generators for life support, and a high-voltage electrostatic precipitator to kill off any bugs. The latter acted like a nasal filter beneath which lay the Elephant Snot Tank, a drip tray that collected the horrible grey slime of our corporate mucus. It was an amazingly healthy system! Bugs didn't stand a chance. The precipitator wiped them out. What a pity it is that passenger aircraft don't have similar systems.

Jim and I then continued through the lead-lined tunnel festooned with an immaculate array of bright, stainless steel, reactor system pipework. At the far end were two portholes through which the unmanned Reactor Compartment could be inspected when the reactor was critical. Then we passed through the after tunnel door into the hidden world of the Back Afties. What a shock! It was filled with acrid blue smoke. I felt a mild sense of panic.

'Are we on fire?' I asked.

'No,' said Jim nonchalantly. 'It's just the lagging. Hydraulic oil drips on to it during maintenance and burns off when we warm up. We're flashing up.'

I shuddered to think of what was going into my lungs.

There was barely enough room to swing a cat back aft. Every square inch seemed to have been filled with heavy machinery, pipework, and grey electrical control boxes, with rat runs in between for the operators. It had the appearance of a scrapyard. Sections of the deck were missing and temporary rails had been rigged overhead, from which a heavy pump was being lowered by chains into the lower level of the turbo-generator room. If simulator training had been nerve-wracking, this was terrifying. I would have to become thoroughly familiar with every valve, switch, alarm, grey box, pipe and pump. And would all this dismantled kit work correctly when put back together?

We stuck our heads into the Manoeuvring Room. It was identical to the simulator at Dounreay except for the baby alarm beside the Engineer Officer of the Watch's stool.

'What's with the baby alarm?' I asked.

'Ah,' said Jim, 'that's a rabbit (an unofficial fit). I rigged it for having private chats with the Officer of the Watch for'ard.'

When we reached the Engine Room ladder, we ducked down two decks to Lub Oil Alley in the bowels of the boat, just below the massive gearbox. This was where the main lubricating oil pump lived. Down there, the smell was entirely different. It was like breathing the odour from a can of engine oil.

'What's the teapot for?'

'Priming the bilge pump.'

'A teapot for priming a pump!' I exclaimed. 'They didn't mention that at Greenwich.'

The *piece de resistance* was the en-suite toilet, a filter funnel known as a 'pig's ear' stuck on to the end of a pipe that took unwanted piddle down into the main bilge tank. It looked remarkably like a voice pipe but the two should never be confused – voice pipes end at someone's ear.

I had imagined that the engine room of a nuclear submarine would be pristine, hi-tech, and Space Age. In fact, only the reactor was hi-tech. The steam it generated, so called saturated steam, was fed into machinery that was Second World War technology with all the traditional engineering problems that entailed: bilge water, steam leaks, pump priming, inaccessible spaces etc. The reactor, sitting in its unmanned ivory tower, merely boiled water. It was the world's most expensive kettle.

Preparing a nuclear-powered submarine for sea is a two-day operation that begins with comprehensive pre-critical checks. These involve double-checking that every single reactor-related valve both inside and outside the Reactor Compartment is in the correct position and that all electronic monitoring and protection systems are tested and working correctly. The Protection System is designed to operate on a fail-safe principle based on three criteria: sensitivity, integrity and responsibility. The reactor monitoring sensors had to be sensitive and reliable enough to detect change but a submarine cannot afford to suffer automatic shut down of its power plant just because one of its sensors becomes defective as it could hazard the boat. So there were three identical 'guard lines' and the reactor would scram automatically only if two or more indicated the same problem which gave the system responsibility. It was called two-out-of-three logic. (Nowadays, three-out-of-four is used).

After pre-critical checks, the Reactor Compartment plug is shut and thereafter the compartment remains unmanned, because the radiation levels inside would be fatal to humans. The reactor control rods are then drawn out slowly until criticality is achieved. This is the nuclear tipping point. Continued rod withdrawal beyond that would lead to super-criticality, a massive uncontrolled surge of nuclear power. That is what happened when Chernobyl went off the clock. Our reactors though were self-regulating and would self correct as the rising temperature made the reactor less reactive – engineering brilliance. Chernobyl had an inherently dangerous design characteristic and the Soviet operators caused that disaster through bypassing safety systems. Our safety management would never have allowed that.

Once the reactor was critical, the steam system had to be warmed through in the time-honoured fashion, before the turbines and propeller could be turned. It is a very slow and painstaking process. Unlike a jet fighter, a nuclear submarine cannot be scrambled.

Over the next six weeks, I shadowed Jumping Jim whilst studying like mad for my Fleet Board exam. Despite twelve months of hard study ashore, there was so much more still to learn. I now had access to the commendably readable Secret Atomic books, the bible for the reactor plant. I also had to read Standard Operating Procedures, which gave step-by-step details on how to do everything from starting the reactor to connecting shore supplies. Then there was the sacred red book of Emergency Operating Procedures which had to be memorised. It gave me the drills for dealing with every possible plant emergency from major steam leak to reactor scram (an unintended shut down). There were also Ship's Standing Orders, pipework and electrical system diagrams, and the location of key valves and switches to memorise, plus the latest Design Authority memoranda. I was spending eight hours a day on watch and the rest studying.

Six weeks later, I was nestling in Paul Anthony's small cabin having a *viva voce* examination by Cambridge-educated Commander 'Neddy' Purvis, the Squadron Commander (E) who would later rise to Navy Board membership as Chief of Fleet Support and a knighthood. I passed, seemingly without difficulty.

The next thing I knew, I was on watch on my own as a fully qualified Nuclear Engineer. Discounting my time in diesel boats and torpedo trials, it had taken eight years of training to reach this stage. I had also accumulated eight years seniority as a Lieutenant and been promoted to Lieutenant Commander. I had all the trappings of a Senior Nuclear Engineer but was a complete novice.

As luck had it, in my first solo watch I had to take the boat into the Full Power State, the first time we had done that since I joined. It was exhilarating. The reactor's main coolant pumps were switched to fast speed and the large turbo-driven feed pumps were flashed-up to ensure a plentiful supply of feedwater to what would become very thirsty boilers. We were doing noise ranging – measuring the boat's radiated noise levels in all our various machinery states – and much of the watch was spent doing high-speed runs with U-turns at either end. In a high-speed turn, a nuclear submarine leans into the turn like an aeroplane. The hull design was so perfect that even at maximum speed, there was not even the slightest vibration.

The magic of our reactor design now became fully apparent. The slightest touch on a tiny joystick would take five thousand tons of submarine from crawling speed to thirty miles-per-hour in a matter of seconds. The reactor simply generated the power on demand. If we went to Stop, it similarly reduced its power without human intervention. It was what was called load following and self regulating – pure dead brilliant.

Other aspects of life on board were more familiar. One morning after communing with nature, the Heads Flushing Valve jammed fully open and a continuous flow of fresh Atlantic seawater came pouring into my lavatory pan. To empty the fast filling bowl, I had to operate a flap valve with my left foot. I did, but only to discover that the U-bend was blocked. As the bowl filled with water, my 'morning George' began to rise inexorably towards the lip of the pan. I reached for the toilet brush and began frantically ramming it in and out of the U-bend but to no avail. The bowl continued to fill.

I now had my right hand trying to shut the flushing valve, my left foot pumping the pedal and my left hand manically poking the toilet bowl with a brush. My fourth limb was required for standing on. I was fully committed but the level in the pan continued to rise. It was clearly going to overflow. My precious offering was about to be washed away into the adjacent Control Room where it would trigger a flood alarm as well as creating a stink. The whole submarine would go to Emergency Stations all because of my errant turd.

As I had to remain at my post like Boy Seaman Cornwell VC at his gun turret, I could think of nothing else to do but shout: 'Help.' My God, did I feel stupid. 'Help' is not a cry that features in the lexicon of submarine emergencies.

As luck had it, the Officers' Heads were next to the Captain's cabin and he was first to respond. 'What the hell's going on?' he roared, sticking his head into my toilet cubicle. He probably thought I wasn't potty trained. I felt utterly pathetic. Thank goodness I had pulled my trousers up.

At that moment, a pair of hands came through between my legs and a voice called: 'Don't worry, Sir. I'll catch it.'

It was the Outside Wrecker, the man responsible for all mechanical systems forward of the reactor, including the heads.

****

When it was *Conqueror's* turn to fire a *Tigerfish*, I was back on home ground. As the trials team had come on board the night before and rigged their data recorders, I could not resist a practical joke. I was very partial to liquorice and kept a box of bootlace liquorices in my drawer. During the night, I carefully replaced the thin black wire for the data recorder at the Torpedo Guidance Unit with bootlace liquorice and re-ran the real wire out of sight. Nothing had been disconnected.

The following morning, when we reached the last ten seconds of the countdown, I was standing opposite Chris Meyer who was conducting the firing. He had just reached 'Five-Four-Three-Two-' when I reached across him.

'What the hell is this?' I exclaimed, ripping the liquorice off the bulkhead and stuffing it into my mouth.

'-One-Fire!' he called, jaw dropping in disbelief. He had just launched our precious *Tigerfish* on a recorded run and I had eaten the recorder wire.

Submarines are frequently involved in scientific trials, usually to do with developing new equipment. *Conqueror* was to investigate bioluminescence caused by tiny sea creatures called phytoplankton and dynoflagellates; I christened them fighter-pilots and flagellators. These tiny creatures give off light when agitated, rather like Jumping Jim. I was very familiar with this phosphorescent phenomenon from sailing on the West Coast of Scotland but had not known exactly what caused it. Our scientists had recognised that bioluminescence could indicate if a submarine had passed by. It could, they thought, be a new method of submarine detection. Accordingly, *Conqueror* was loaded with seawater sampling devices and complex biological analysers and off we went.

The poor scientists had no luck at all; there was simply nothing for them to analyse. Until one night when I knew for certain that bioluminescence was present and went along to congratulate them.

'You must be having a field day,' I said.

'Why?' they asked.

'Plenty of bioluminescence now.'

They looked puzzled. 'How do you know? We're getting nothing on the analysers.'

'I've just flushed the heads. The pan was full of it.'

Could we now detect enemy submarines through our lavatory pans?

****

We then deployed for a three-month secret patrol in the Mediterranean. Unlike in *Otter* when the long passage was on the surface, *Conqueror* did a fast, deep transit – twice as fast as *Otter's* top speed – and we remained submerged all the way from Faslane to the far end of the Mediterranean and back again to La Spezia in Italy, a distance of some five thousand nautical miles without surfacing. Our mission was to work as forward sweeper for the US 6th Fleet, based in Naples. A Soviet submarine was thought to be on the prowl and our job was to flush it out. We did. We sanitised the Eastern Mediterranean with our high-powered active sonar and found him, just as *Valiant* had found *Andrew* in the exercise some years previously. The Americans were suitably impressed: 'You goosed him!' the American Admiral signalled. 'There's a crate of champagne for you on my piano.'

One afternoon during that fast transit, I was heading aft to take the watch when there was a loud grinding noise and the submarine lurched violently to starboard. It was an alarming moment. I felt sure we had grazed against the Soviet and the difference between grazing and a fatal head on collision is miniscule. I had hardly taken over the watch when the hotline from the Control Room squealed. It was one of the old hand-wound telephone alerts and one could tell from the noise how hard it was being wound at the other end. The winding was manic. I answered it.

'Stop that bloody noise immediately!' It was the Captain. He was furious. Submariners are paranoid about noise.

'Yes, Sir,' I replied. 'Which noise?' The machinery spaces were full of noise.

'Bloody well find out!'

'Yes, Sir.'

I left my Chief in charge of the watch, took an electronic stethoscope and set off on a mad round of the machinery spaces, listening to each machine in turn like a demented doctor. I had never done this before and was not prepared for my findings. The main gearbox sounded like a bag of cutlery in a washing machine; the clutch was clearly about to collapse and the thrust block, where the propeller drive is transmitted to the hull, sounded like it had wiped its bearings. Everything sounded as if it were falling apart.

Then I stuck the stethoscope against the pressure hull at the stern in an attempt to hear the propeller. Eureka! It sounded like someone was running a ruler along a radiator. This was no rattle. We had something wrapped round the propeller and that was very bad news. We had only one propeller and without it, we would be up the proverbial creek without a paddle.

'We've got something round the propeller,' I reported to the Captain.

'No, we have not! Now go and sort out that bloody noise.'

My *Otter* experience kicked in. I was not going to be overridden when I knew I was right.

'There's no doubt about it. The noise is coming from the propeller.'

'Hmmm.'

We went up to periscope depth and stuck our tail in the air like a duck. The propeller broke the surface. I was right. There was a steel wire rope wrapped round it. We gave a quick burst of astern revs and held our breaths. As a diver, I had wrestled with ropes and fishing lines wrapped round yacht propellers and knew very well they could jam so tightly it would almost become a solid mass. Miraculously, our wire was simply thrown off. We heaved a collective sigh of relief. The tail cone of the propeller on a nuclear submarine was so beautifully designed that it had not been possible for the wire to jam between it and the submarine's hull. It had merely been wound round the blades like thread on a bobbin and had unwound with equal ease. Problem solved. My credibility with the Captain hit new heights.

The degree of stress and exhaustion on board became evident from nocturnal events in the eight-berth cabin. I had the top bunk of three and was lying awake one night when my bunk gave a heave. It felt like a minor seismic event. The earth had literally moved for me. Then it moved again and again with increasing force. The seismic event was moving up the Richter scale.

My first thought was that I was about to be tossed out of my bunk whilst still cocooned in my sleeping bag and would fall on top of the cabin chair and break my back. I reached for the grab bar above the bunk. As I did so, the heavy bunk gave one mighty heave, came out of its sockets and fell down on top of John Board in the bunk below. I was then left hanging from the grab bar in my sleeping bag, knees up to my chest, like a giant bat. Now I was stuck. If I lowered myself, it would be on top of John who was already being crushed under the weight of my bunk. I felt guilty but could not think why. Mine was the only bunk to have collapsed. What would my cabin mates think?

Someone switched on the cabin lights and the others came to my rescue. Together we raised my bunk off John and refitted it into its sockets. When we tried to revive him, he woke with a start. He had slept through the entire incident.

A few days later, it happened again. This time I was lying in my bunk reading a book. It was during the afternoon and the cabin lights were on. I reached immediately for the grab bar but this time, peered over the edge. Below me, John was clearly having a nightmare. His knees were up against the underside of my bunk and he seemed to be trying to heave it off him as if it were crushing him. With one mighty heave, he raised it out of its sockets and it collapsed on top of him once again, leaving me swinging from the grab bar. Now I understood the seismic phenomenon – John seemed to be suffering from some form of subconscious claustrophobia.

On another occasion, when I came off the Middle Watch (00.00 – 04.00) and entered the cabin, Graham Stevens whipped back his bunk curtain, stabbed the bulkhead with a pointing finger and shouted at me: 'You're supposed to be watching these dials,' He was looking straight at me, eyes tight shut.

We also had a guest Commander sharing the eight-berth in preparation for taking Command of his own nuclear boat. One night, I heard him cry out in a loud wailing voice: 'No No No,' followed by 'No No,' followed by 'No,' and then 'Yes.' I shall never know whether he had been trapped in some subconscious erotic compromise or was merely having an argument with his wife.

Despite my private self-doubts, I seemed to be one of the least troubled officers.

On arrival in La Spezia, I noted a deep chinstrap-like gouge round the leading edge of the fin, which I deduced was because of the steel wire rope getting caught. Five thousand tons of submarine travelling at twenty-five knots must have pulled the wire bar taut and the propeller had then chopped it. One shudders to think of what happened at the other end of the wire.

****

The exercise of stethoscoping our machinery opened my eyes, or at least my ears. Underwater warfare is all about sound. Target noise is what sonar and homing torpedoes detect. The whole aim of underwater warfare is to hear the target before it hears you – that's called detection advantage. Then one must be able to guide a homing torpedo in for the kill, ideally whilst your victim is still blissfully unaware that he is under attack. No Marquis of Queensberry rules here.

As part of that equation, it is vital that one's own submarine is ultra quiet. This is so the target cannot hear you and also to prevent deafening your sonar systems with your own machinery. Whales solved this latter problem eons ago by isolating their hearing organs from their skull bones. Thus, a Killer Whale can still hear while it is crunching a seal. (Try munching potato crisps with your hands over your ears).

Armed with my electronic stethoscope, I set about systematically monitoring all our machinery and was thus able to advise the Captain on how to make *Conqueror* quieter. That impressed him. As a bonus, my listening regime also revealed the health of our machines and that impressed the Engineers. (Hearing a new noise in the car is often the first sign of an impending problem). I had just illustrated one of the first steps in what was to become the new science of Signature Reduction, also known as Stealth Technology.

When dived, a submarine has no visible shape. It has only a noise signature which is how it is detected and identified as a friend or foe. Capturing the noise signatures of Soviet ships and submarines was one of the main purposes of submarine intelligence gathering during the Cold War. It was akin to the voice recognition technology used in telephone tapping. Our submarines carried libraries of Soviet signatures gathered by sneaky boats and our on board computers analysed all incoming sounds. It is exactly how whales and dolphins behave. They rely on sound for navigation, recognition, hunting, and communication.

When a dolphin looks at another dolphin underwater, it does not see the dolphin shape we recognise. It scans it with ultrasound and sees only the bits that reflect sound, such as air trapped in the lungs or the skeleton. When a dolphin looks at a pregnant female, it will see the unborn calf in the womb. It is conducting an ultrasound scan, just as hospital scanners do with pregnant women. A dolphin midwife, and there are such things, is able to watch the calf leaving the womb.

On that Mediterranean patrol, I had the great pleasure of watching a school of dolphins from beneath through the periscope. I had often heard them cackling away on the underwater telephone but had never before seen them from this angle. They chatter the whole time.

The standard naval priorities are *float* – *move* – *fight* in that order. Traditionally, the Engineers are responsible for *float* and *move* and the Seamen for *fight*. However, my noise project now engaged the Mechanical Engineers in the *fight* bit because, in acoustic terms, we were indeed 'all of one company.' Based on my initiative, the 'Clankies' would in future inherit responsibility for Signature Reduction.

****

During the patrol, I also appointed myself as Entertainment Officer and organized, amongst other things, a poetry competition. It was possibly the world's first underwater poetry competition. I wondered if there would be any entries and expected at most, a few smutty limericks. How wrong I was. When doing rounds of the turbo-generator room one night, I found a Leading Mechanic sitting on top of the main bulge pump scribbling his entry.

And in the Sick Bay, I found our Petty Officer Medical Assistant composing a most serious poem about the birth of his first child.

What a revelation.

On competition night, the Junior Rates mess was full and Chris Lowe, a theatrical type, read out all the entries. A vote was then taken. I was the only one to submit a smutty limerick. One lives and learns.

****

Things were going extremely well for me. I felt thoroughly accepted by my brother Engineers, by the Seaman Officers, by the Ship's Company and, most importantly, by the Captain. I had found my feet in the nuclear world. It would have been ideal had I been able to remain in *Conqueror* but life in the Navy is never that simple. In the mail at Gibraltar, I received my next appointment. I was to become senior Electrical Officer in *Courageous*, a plum appointment as she was engaged in spying on the Soviets. This was clear evidence that the Captain had given me a good report and that the Squadron had faith in me as a nuclear engineer. But every cloud has a lead lining. *Courageous* was nearing the end of her commission and would be going into Chatham dockyard for a two-year refit. My main role would be to take her through the refit, not play spies in the Barents Sea.

My heart sank. Chatham was only a stone's throw from Greenwich and I had just moved Kate and the family back from there to Faslane, five hundred miles away. Richard had settled into his new school and we had bought a house. I had asked for an operational boat running from Faslane, not one being overhauled in Chatham. Besides, after two years I would have to make yet another move as submarines did not operate from Chatham.

It was time to face the inescapable fact that I could not provide Kate and the family with a stable home whilst pursuing a career in the Navy. Kate had not joined the Navy; she had married me. She was too loyal to have complained but I owed her a life, which remained my most important duty. I had to make a choice, family or career. Rationally, it was easy – Kate came first and always would. Emotionally, it was agonising. My stomach churned. A career in the Navy had been my ambition since earliest childhood. Now I had made it. I was living the dream. I was so intensely proud of being a fully qualified nuclear submarine engineer but the time had come to take some control over my own destiny. With the heaviest of hearts, I tendered my resignation. I would revert to my secondary ambition of becoming a schoolteacher and plough back what my wonderful teachers in Coatbridge High School had given to me. It seemed like a perfectly honourable course of action but I felt like a traitor.

On return to Faslane, *Courageous* lay alongside the jetty next to us. At the reception to welcome us home, I met Chris Childs, her Senior Engineer. He would be my new partner-in-crime and cabin mate. As we had joined the Navy together and played in both the Dartmouth and Manadon football teams, we were old friends and he was clearly looking forward to my joining. Immediately, he hauled me over to meet my Captain-to-be.

'This is our new Electrical Officer, Sir,' he proclaimed, as if he had just pulled off a major recruiting coup.

My Captain-to-be was tall, serious and red-haired. He turned and gave me a distasteful look: 'I hear you've resigned.'

The remark struck me like a dagger through the heart. 'I'm afraid I have,' I replied, 'but I shall be doing my very best until the day I leave.'

'Hmmm!' he retorted with contempt. 'They all say that. Rotten apples.' With that he turned his back.

I was stunned. He saw me as a poisonous influence, a bad smell, as a dropout lacking commitment. He had swatted me like a fly. For the second time in my career, I felt alienated before I had joined. Now I faced the prospect of serving under a Captain who was openly hostile to me.

Memories of *Otter* came flooding back. This was so horribly *deja vu* and I had vowed never to let it happen again. If I did, I would be on a hiding to nothing. I would have enough on my plate simply trying to fit in as the new boy in a fully worked up team. I did not need pre-conceived hostility from the Captain as well. I left the reception and went back on board *Conqueror*. I had still to acknowledge my appointment. Now I would.

Acknowledgement of one's appointment in the Royal Navy is a highly stylised ritual in which one writes in manuscript to one's future Commanding Officer in pen and ink using the standard format: *Sir, I have the honour to acknowledge my appointment to Her Majesty's Ship...under your Command.* And one signs off with: *I have the honour to be, Sir, your obedient servant.* On this occasion, I went freelance and added an extra paragraph acknowledging that I had been classed as a rotten apple and requesting further instructions. I was cutting my own Naval throat but so what? I had already tendered my resignation. I had nothing to lose.

I had also vowed never again to let a personal letter of mine be torn up and thrown into a bin, so I went on board *Courageous*, legs like jelly, found the Correspondence Officer, and told him to ensure that my letter was logged in officially. I had, in effect, rendered my appointment untenable.

It was probably the first time in the history of the Royal Navy that a junior officer had written such a letter to his appointed Commanding Officer. It was tantamount to refusing to obey an order. In Nelson's day, I would have been hanged from the yardarm. The authorities would not be amused. I would be branded as bolshie, as a member of the awkward squad, but, to quote King Lear, I felt *more sinned against than sinning.*

My Appointer must have convulsed on hearing the news. The following morning, he phoned me at home. He was rip-roaring, seething mad. 'Thompson,' he declared, 'you are *persona non grata* in the Third Submarine Squadron. I'm transferring you with immediate effect to the Tenth Squadron as a spare officer. You can stay there till your time is up. You've made yourself completely unappointable.'

No doubt about it now, my career was well and truly over.

CHAPTER 14

Fire Down Below

'And should you like to fall into that pit, and to be burning there for ever?'

CHARLOTTE BRONTE

I felt like a condemned man. I had eighteen months notice to serve and then would be dead as far as the Navy was concerned, but it was still my Navy and I would give it my best until the end. The Tenth Squadron was the Polaris squadron. I was courteously received but had no job. I was what was known as the Shitty Little Jobs Officer.

On 2nd May 1976, I was at home in bed listening to the radio. As the chimes of Big Ben receded, the midnight News opened with the headline that there had been a major fire in a British nuclear submarine. My heart skipped a beat. Whichever boat it was, I would have friends on board. It was *Warspite*, sister ship to *Conqueror*, in the Third Squadron. She was on a courtesy visit to Liverpool to mark the anniversary of the Battle of the Atlantic. Her Senior Engineer was Tim Cannon who had joined the Navy with me and had been in my submarine training class.

In a flash, I was out of bed, throwing overalls, pyjamas and a toothbrush into a bag and heading for Liverpool. *Persona non grata* I may have been but Tim would need help and I knew I could be of assistance in some way or other, no matter how menial. As the roads were empty at that time of night, I drove like a maniac and four hours later was in Seaforth Docks in Liverpool. There I found a surreal scene. The jetty was floodlit, there were lines of fire engines, blue flashers still going, a snakes' wedding of fire hoses, a loose crowd of sailors, policemen and firemen, and a pair of large, portable, diesel generators which should have been chugging away but were silent. It felt as if I had just strayed on to the set of a disaster movie. Alongside the jetty, *Warspite* looked perfectly normal.

As I descended through her accommodation space hatch, I knew immediately that things were very far from normal. The atmosphere was thick with the acrid smell of burning rubber and stale chemical smoke, and the boat was virtually in darkness with only dim emergency lighting for illumination. It was like entering a coalmine.

Tim and David Mattick, the Electrical Officer, were standing in the Wardroom in filthy white overalls looking dazed and exhausted. The fire had started around four o'clock the previous afternoon and had raged for four-and-a-half hours. Neither of them had slept. For twelve hours they had been wrestling with the worst emergency in the history of the British nuclear submarine programme.

On the boat's arrival in Liverpool, the reactor had been safely shut down and the submarine had switched to shore power supplied by the portable generators on the jetty, a standard procedure in non-naval ports and therein lay the source of the problem. When the portable generators failed, the submarine lost its electrical supply and the duty watch had to start the on-board diesel. Disasters rarely happen for a single reason and on this occasion, the boat's diesel then suffered a pressurised oil leak on to its hot exhaust pipe. Instantly, that started an inferno creating intense heat and volumes of dense black smoke. In seconds, the machinery spaces had filled with toxic fumes and to survive, the watch-keepers had been forced to plug into the nearest breathing nipple in the Emergency Breathing System (EBS) that ran throughout the boat. They could only escape if they managed to find their way from nipple to nipple. In the smoke and blackout, they could not – they were trapped

Tim's first concern had been for the safety of his trapped watch-keepers, coupled with concerns for the safety of the reactor. Then there was the actual fighting of the fire, a terrifying physical challenge undertaken in blinding smoke, intense heat, and a claustrophobic environment, whilst wearing breathing apparatus. To make matters worse, with all electrical supplies lost, they had been fighting in the dark. When electrical power was lost, the reactor coolant pumps had stopped and the reactor (already shut down) had automatically failed-safe into Emergency Cooling (a static heat exchanger).

That left two most serious questions to address. First, should a Nuclear Accident be declared and if so, should the submarine be sunk to reduce any potential radiation problem? This would trigger Liverpool's Emergency Plan for a major disaster, cause widespread public panic, and have profound political implications for the entire British nuclear submarine programme. Second, there were the men trapped back aft who had to be saved. Cool heads were required. Tim bore responsibility for such decisions.

Leading from the front, he had donned breathing apparatus and gone back aft into the inferno to lead the trapped men to safety. Thanks to his outstanding courage, the watch-keepers were saved though hospitalised due to smoke inhalation. When the Liverpool Fire Brigade arrived, Tim and his team had led them down into the death traps of unfamiliar machinery spaces and together, for several hours, they had fought the fire. For their bravery, Tim and two of his men (POMEM Middleton and MEA Ashcroft) were awarded Queen's Gallantry Medals, with three others being Mentioned in Despatches.

The evacuation of the watch-keepers had left no one to monitor the reactor state which exposed a design omission: there was no duplicate reactor instrumentation forward of the Reactor Compartment. In any case, electrical supplies to the instrumentation had been lost. The state of the reactor could no longer be monitored, an unnerving situation for any nuclear engineer. However, manual radiation monitoring both inside the boat and externally on the hull indicated that its state was perfectly normal.

When I went back aft, I could scarcely believe my eyes. Everything was soot-blackened and much of the electrical wiring, despite its fire resistant properties, had clearly been burning. The stench of burnt rubber is engraved in my memory. The immediate requirement was to survey the key wiring and restore as many supplies as possible, in particular supplies to reactor instrumentation. I was in the right place at the right time. At eight o'clock in the morning, I phoned the Squadron office to let them know that their Shitty Little Jobs Officer was otherwise engaged.

When *Warspite* had been made safe, she was towed to Devonport Dockyard for a major overhaul and, two years later, emerged as good as new.

There are very few ports in the world prepared to accept nuclear-powered vessels. Before they can, they must have in place a public safety plan for dealing with a potential Nuclear Accident, have a jetty that is remote from the population centre, and must have conducted a radiological survey of the area beforehand to ensure it would not impact on the local radiological environment. As *Warspite's* reactor had remained safe throughout the fire, no changes to these rules were necessary and nuclear submarines continue to visit Liverpool.

****

Fire is one of the worst things that can happen in a submarine. It very rapidly burns up the limited amount of oxygen on board and turns it into deadly carbon monoxide and suffocating carbon dioxide whilst also filling the boat with toxic smoke. The crew must plug immediately into Emergency Breathing System connections to survive and that tethers them to the nearest breathing nipple. The fire has then to be fought by men in breathing apparatus and cannot be fought simply by pouring tons of water on to it through fireman's hoses like in a building fire. That could sink the boat, cause major electrical failures, scram the reactor, and leave the boat without the propulsion required to reach the surface.

Had *Warspite's* fire happened at sea, the boat would have surfaced for fresh air and safety, but there would have been limited chance of escape. A nuclear submarine has no upper deck, no lifeboats, and scarcely enough room for the crew on the casing. If the sea were to be anything other than calm, they would have been washed overboard. Fire is a submariner's nightmare.

Warspite survived, but the Soviet nuclear-powered, nuclear-missile-carrying submarine, *K278 Komsomolets*, did not. On 7th April 1989, she suffered a major fire on patrol. Like the ill-fated *K19 Widowmaker* before her, she had deployed on her first operational patrol and was running submerged at a depth of 1,099 feet (335 m), one hundred miles South West of Bear Island in the Barents Sea. The fire broke out in her engine room due to an electrical short-circuit and was so intense that it melted bulkhead glands. When rubber seals melted on high-pressure air systems, the fire was then fanned up by this fresh supply of oxygen and further fuelled by leaking diesel and hydraulic oil, which created the perfect inferno. The heat was so intense that it melted the cable penetration glands in the pressure hull, thus creating holes through the hull that sealed her fate. Holes in the hull cannot be shut off with valves. She was condemned to sink.

A mere eleven minutes after the fire had begun, with all power lost, her ballast tanks were blown in emergency and she shot to the surface out of control. Her crew were then faced with the choice of death by burning, drowning or freezing. Most abandoned ship but the Captain and some other brave souls remained on board until she sank five hours later, long before any rescue craft arrived. Five of the latter managed to escape in a small survival capsule but when that too sank in rough seas, only one survived. Out of her crew of sixty-nine, forty-two lost their lives, most dying from hypothermia or drowning in the ice cold water. A nearby fish factory ship saved the remainder.

Komsomolets was a fully manned, well-designed, modern, operational submarine that had recently achieved the submergence record of 3,350 feet (1,020 m), far deeper than any British or American boat could dive. But fire respects nothing. It took only eleven minutes to destroy her. The hulk now rests one mile down at the bottom of the Barents Sea.

The Soviet Union never admitted such failures as the Party was not prepared to lose face. The reason for details of the *Komsomolets* disaster being known is that, one year earlier, Mikhail Gorbachev had become Soviet Head of State and had introduced *glasnost*, a new regime of openness. Gorbachev authorised release of the *Komsomolets* investigation report and I was able to read it. As with the *K19* disaster, the salient feature was the heroism of the Russian submariners, a number of them helping shipmates to safety but perishing in the process. When I read that report, my stomach churned. I identified totally with those poor, brave, Russian submariners. They were men just like me, faithfully serving their country. I could imagine their plight with frightening clarity.

Eleven years later, just after the Cold War ended, the Russian Navy also lost *Kursk*, the pride of their fleet, with all one hundred and twenty-nine of her crew. This time, highly volatile peroxide torpedo fuel had accidentally mixed with its matching kerosene and caused first an explosion and then a fire in the torpedo compartment. That, in turn, caused the simultaneous detonation of the warheads in the other torpedoes. The resultant explosion was of such ferocity that it blew the front end off the submarine and squashed up all the forward bulkheads like a concertina. The explosion was so enormous that it was detected on seismographs as far away as Alaska.

Miraculously, the engineers on watch aft of the reactor compartment actually survived the explosion but died a slow, miserable death trapped in their icy-cold, pitch black, flooded compartment. Some even had time to scribble farewell messages to their loved ones. Perhaps they could have been saved but the Russian Navy proved completely inept in its rescue efforts and wasted days in an instinctive Soviet-style cover up. When far too late to save life, they did eventually seek help from the NATO navies.

I had always thought it sad that we could not rub noses with our fellow submariners in the Soviet navy. After all, who else in the world has more in common with a submariner than another submariner? Who else spends months on end cooped up in a steel tube suspended under the sea in the service of his country? And submariners, almost by definition, are the most cheerful, amiable and sociable souls on the planet – they have to be. Would it not be wonderful, I used to think, if all the submarines in the Atlantic could agree secretly to rendezvous somewhere and have a thoroughly good run ashore together? Without doubt, it would be one helluva party.

That wish came true in 2001. It was the centenary of our Submarine Service and a multi-national force of guest submarines assembled at Faslane, including a Russian. The end of the Cold War rainbow had been found. We were all friends and that was official. It was uplifting to see young Russian submariners walking freely through the streets of Helensburgh. Though they were penniless, the young Russian sailors had no need to buy a drink, for the generous folk of Helensburgh were always ready to pay. It was reminiscent of my father and his shipmates in America during the Second World War where they had enjoyed the seemingly bottomless pit of American hospitality.

Our Submarine Centenary Banquet was held that same year in the magnificent Lancaster Town Hall. It was presided over by Mike Boyce, term mate, fellow submariner, and by then First Sea Lord. I was given the honour of delivering the main after dinner speech. In the audience, amongst many other nationalities, was the retired captain of a Soviet missile submarine named Ivan who had counter patrolled against us when I was at sea. After dinner, through his glamorous Russian interpreter, I was told that he had thoroughly enjoyed my speech and would like me to join him at the bar.

Elena, his interpreter, was, I felt sure, KGB-trained and licenced to thrill. She was wearing a slinky, body-hugging, long black dress with a décolleté that reached from her chin to her navel, her cleavage being tantalisingly obscured by a black stole which, very deliberately, she would let slip from time to time to maintain my attention. She was the most overtly seductive woman I have ever met. When she invited me to join them, I wobbled at the knees.

At the bar, Ivan and I shook hands. Then he excused himself, returning two minutes later clutching a bottle of best Russian vodka and a photograph album. Along with Commander Jeff Tall OBE and Commodore Martin MacPherson OBE, two veteran British submarine Commanding Officers and personal friends, we demolished the vodka as we viewed Ivan's photograph album. It had photographs of his wife and family, his crews, his submarines, the Soviet submarine memorial in Murmansk, and pictures of Soviet submarine memorial parades. His pride in the Soviet Navy was just as great as ours in the Royal Navy. We were, as I had always known, blood brothers. It was our political leaders who had forced us to be enemies.

****

Some weeks after returning from the stricken *Warspite*, I answered the phone at home unusually by saying: 'Helensburgh lunatic asylum.'

'Is that Lieutenant Commander Thompson?' a voice asked.

'Yes,' I replied.

'It's your Appointer. I'm coming to you cap-in-hand,' he said apologetically. 'We're scraping the barrel for nuclear-qualified Electrical Officers. If I appoint you to *Revenge*, based at Faslane, would you withdraw your resignation?'

That put me on the spot. The fact was that the Navy needed me but it was hardly the most flattering choice of words. I had wanted Fighter Command, not Bomber Command, and *Revenge* was in the latter. She was a Polaris missile submarine.

'Yes.'

CHAPTER 15

The Strategic Nuclear Deterrent

Fearful politicians   Corridors of power

Resolute Defence chiefs   Ready for the hour

Sending coded signal   Penetrating brine

From their secret bunker   Radiates the sign

Fifteen minutes notice   Out the missiles fly

Taking retribution   Through the purple sky

Tinker, tailor, soldier   Burns the human fuel

Burns the little children

Burns the little school.

Atmospheric chaos

Radiation bloom

Diplomatic contacts

Sent from shielded room.

'Have you learned your lesson?'

'Was……..our…….message……clear?'

No communication………………Nothing left to fear.

AUTHOR

Revenge was enormous. She was like *Conqueror* cut in half with an extra section carrying sixteen intercontinental ballistic missiles sandwiched between. As she was carrying the nation's Strategic Nuclear Deterrent, she was berthed in the high security area at Faslane, guarded by armed Royal Marines and Ministry of Defence Police. I was now in the business of Mutually Assured Destruction and it was a different world. This was not about practising for a dogfight with the Soviets; this was to ensure that such a fight would never be necessary.

In that sense, *Revenge* was on the equivalent of a war footing. For her, things could not be more serious. Her role was to prevent a third world war. Were such a war to happen, she would have failed. It was all utterly real and of the utmost importance for the human race – and it reached to the very core of the nation. We had bought the Polaris missile system from the Americans but the nuclear warheads and everything else were British. As a country, we had the power to launch our own nuclear retaliation.

Normally, British nuclear weapons are assigned to NATO, our targets being set in accordance with NATO targeting plans to prevent duplication of UK and US counter-strikes. The scenario for unilateral British action was therefore difficult to imagine: NATO would have collapsed, the Americans would have opted out of European defence, the Soviet Union would have launched a pre-emptive nuclear strike on Britain, with London and several other major British cities being wiped out and millions of British citizens killed. As a result, Britain would have descended into ungoverned anarchy, as depicted in the 1965 BBC documentary *The War Game*, the one Parliament banned on the grounds that it was 'too horrifying for broadcasting.' (The BBC did not broadcast it until 1985 for the fortieth anniversary of the Hiroshima bombing). Such a doomsday scenario had to be avoided at all costs but could not be prevented simply by asking the Soviets to be nice to us. It could be prevented only by the assured threat of similar destruction being wreaked on them.

There was, however, another factor in this gruesome equation. In the Second World War, the Soviets lost twenty million people yet survived to win a convincing victory. If we were to kill only ten million of their citizens in tit-for-tat retaliation, they had already shown they could stomach double that loss and survive; they had a population of 250 million. For Britain to lose ten million people at a stroke would be unimaginably catastrophic. In the Second World War, we lost fewer than half-a-million. Independent nuclear retaliation would therefore be our final act in any Third World War. It would not be the overture for a longer conflict.

For it to be a deterrent, the Soviets needed to know that our weapons worked and that we really could use them if necessary. They knew. They monitored all our test firings at Cape Canaveral and maintained a spy trawler (AGI) between Malin Head and the Mull of Kintyre to count our SSBNs in and out. On one occasion, the Captain of the AGI at Canaveral, spying on *Repulse's* test firings, actually called *Repulse* by name (Commander R. C. Seaward OBE RN) on VHF to congratulate him on his four successful launches. The Soviets fully comprehended our ability to retaliate. As a result, they had built Anti-Ballistic Missile (ABM) defences around Moscow. To defeat these, the UK had secretly developed *Chevaline*, a highly effective decoy system, and the US developed MIRVs (Multiple Independently Targetable Re-entry Vehicles). Thus the Soviets had no guarantee of safety behind their anti-missile defences.

26th September 1943. My father's ship, HMS *Intrepid*, sinks after a German bomber attack in Leros. I was in my mother's womb at the time. (Leros is the fictional island of Navarone in Alistair Maclean's novel, 'The Guns of Navarone.')
Source credit: *Courtesy of the War Museum, Leros*

16th July 1945. The Manhattan Project detonates the world's first nuclear weapon; the Nuclear Age has begun. The world would never be the same again. Nuclear weapons cannot be disinvented; like them or not, they are here to stay.
Source credit: *Manhattan Project*

17th March 1959. USS *Skate* becomes the first submarine to surface at the North Pole. In August 1958, USS *Nautilus* had become the first 'ship' in history to sail from the Pacific to the Atlantic under the Arctic icecap.
Source credit: *Copyright US Navy*

Maps can be deceptive; a look at the globe reveals the strategic importance of the Arctic Ocean. Prior to the advent of the nuclear submarine, the North Atlantic had been the maritime battlefield. The Arctic would become a front line in the Cold War; it is the meat in the sandwich between Russia and North America.
Source credit: *Private*

Britannia Royal Naval College, Dartmouth, home of officer training in the Royal Navy. It is also where Her Majesty Queen Elizabeth was introduced to her future husband, Prince Philip of Greece, a student at the College. Their introduction was in the Captain's house to the right of the main building.
Source credit: *Crown copyright*

June 1962. Her Majesty Queen Elizabeth arrives to take the salute at my Passing Out Parade. The Cadet guard can be seen on the parade ground below.
Source credit: *Crown copyright*

1962. HMS *Barrosa* at anchor in the Great Bitter Lake on the Suez Canal, en route for joining the British Far East Fleet. The bedstead aerial on top of her mast was the latest in long-range aircraft detection radar. Her role was to operate well ahead of a carrier battle group and provide early warning of attack. I served my year as a Midshipman in *Barrosa*.
Source credit: *Private*

1963. My carrier training was in HMS *Hermes* in the Far East. She carried a nuclear weapon ready to be delivered by a Fleet Air Arm fighter-bomber. The large box-like thing on top of her island is a state-of-the-art aircraft direction radar. Twenty years later, she became famous as Admiral Woodward's flagship in the Falklands War.
Source credit: *Crown copyright*

The Peace monument in Nagasaki, as sketched for my Midshipman's journal in 1962. It sits at ground zero, exactly below where the atomic bomb exploded on 9th August 1945. The right hand points to the sky where the airburst detonation happened. The left hand symbolises peace on earth. I saw no evidence of the nuclear devastation caused seventeen years earlier.
Source credit: *Own sketch*

The Royal Naval Engineering College, Manadon, Plymouth, where Engineer Officers were trained (now demolished). The College had a flight of Tiger Moths to enable Air Engineers to qualify as pilots.
Source credit: *Crown copyright*

The author standing by the gun in HMS *Andrew* in 1969. Strange to think that when this photograph was taken, Britain's first Polaris submarine, HMS *Resolution*, was on patrol with an arsenal of intercontinental ballistic missiles.
Source credit: *Private*

An O Class submarine. The O Class was post war designed and ultra-quiet. It was eclipsed by the advent of nuclear-powered submarines but was nevertheless highly effective in many roles. I served as Electrical Officer in *Otter* and *Osiris*.
Source credit: *Crown copyright*

Photo taken during a Submarine Commanding Officers' Qualifying Course, the 'Perisher'. The frigate is doing 28 knots and trying to run down the submarine – a serious game of underwater chicken. This training prepared COs for carrying out attacks such as *Conqueror's* sinking of the *Belgrano* in the Falklands War. It was also to prove they could remain cool under pressure.
Source credit: *Crown copyright*

1970. The Malin Head AGI (Acoustic Intelligence Gatherer) was a Soviet spy ship that kept station between Scotland and Ireland throughout the Cold War. Its role was to monitor US and UK submarine movements in and out of the Clyde.
Source credit: *Crown copyright*

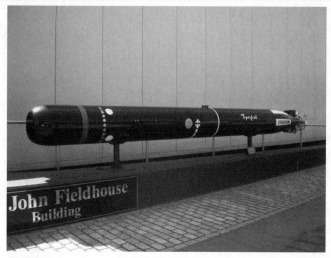

Tigerfish was a sophisticated, long-range, wire-guided, homing torpedo. I spent five years in its development. It was a thing of beauty – if you liked torpedoes. After many trials and tribulations it became the best anti-submarine torpedo in the world.
Source credit: *Private*

1982. This silent assassin, a Valiant Class SSN, was about to test a *Tigerfish* against a decommissioned frigate.
Source credit: *Crown copyright*

Job done. One *Tigerfish* breaks a frigate in two.
Source credit: *Crown copyright*

The Royal Navy Department of Nuclear Science and Technology was housed in Sir Christopher Wren's architectural masterpiece, the Royal Naval College, Greenwich. Somewhat incongruously, there was a small nuclear reactor in its basement.
Source credit: *Private*

The iconic nuclear research reactor at Dounreay on the North Coast of Scotland. Few outsiders were aware that a submarine nuclear propulsion test plant was also located there. It is where Submarine Engineer Officers did their practical nuclear reactor training.
Source credit: *UKAEA copyright*

A nuclear powered hunter-killer submarine slips out of the Gareloch at dusk.
Source credit: *Crown copyright*

A Resolution Class submarine (SSBN) deploying on patrol. She would dive in the Clyde and disappear for 2-3 months. Note that the bow is completely submerged. SSBNs spent very little time on the surface and none at all when deployed. They carried Britain's strategic nuclear deterrent.

Source credit: *Crown copyright*

The author on watch in the Manoeuvring Room of HMS *Revenge*. I was here when the major steam leak occurred. Thanks to my team of highly trained engineers, we survived the emergency and maintained our patrol aims.

Source credit: *Private*

There was very little space in the engine rooms. The head of Leading Mechanic 'Taff' Roden can just be seen amongst the hardware. 'Bungy Mack' risked his life in such a space in an attempt to stop the steam leak. For his bravery, he was awarded the Queen's Gallantry Medal. Source credit: *Private*

The *Revenge* engineers who saved the credibility of Britain's nuclear deterrent programme being awarded Oak Leaves for gallant conduct. (Left to Right: Lt Cdr Paul Thomas (later Admiral); Mechanician McCrum (Reactor Panel operator); Mechanician McDonagh; Captain Andrew Thomson, Captain of the Tenth (Polaris) Squadron; Mechanician Murdoch; the author. Source credit: *Crown copyright*

Leading Mechanic James McWilliams ('Bungy Mack') and me outside Buckingham Palace. He had just been awarded the Queen's Gallantry Medal and I had just received my MBE (Membership of the British Empire).
Source credit: *Crown copyright*

A personal triumph: a Sub Harpoon missile is loaded into the torpedo compartment of HMS *Odin*, much to the annoyance of the Ministry of Defence who had no plans for her to have this weapon. *Odin* deployed with it to the Falklands war zone. She remains the Royal Navy's only missile-firing diesel boat.
Source credit: *Private*

The author with Raffles. This photograph created a spoof story that the Navy was planning to replace sonar operators with dogs trained to recognise Soviet submarines and bark. Dogs would have been much cheaper than sailors.
Source credit: *Crown copyright*

1982. A Sub Harpoon anti-ship missile is fired. Whilst launch compromised the location of the submarine, Sub Harpoon was an 'over the horizon' weapon with a range of 70 miles. To sink *Belgrano* with old Mk 8 torpedoes, HMS *Conqueror* had to close the range to one mile.
Source credit: *Crown copyright*

1993. HMS *Vanguard* arrives in the Gareloch from Vickers Shipbuilders at Barrow-in-Furness. She was twice the size of a Polaris boat and more than twice as lethal. Conceived and built during the Cold War, *Vanguard* entered service after the Cold War had ended, adding fresh controversy over the need for Britain to have an independent nuclear deterrent.
Source credit: *Crown copyright*

1964. When I first arrived at the Clyde Submarine Base at Faslane, it was no more than a depot ship, HMS *Maidstone*, and home only to the 3rd Submarine Squadron, consisting of diesel boats plus HMS *Dreadnought*, Britain's first nuclear-powered submarine.
Source credit: *Crown copyright*

1998. Despite Defence cuts following the end of the Cold War, the submarine base at Faslane continued to expand. By the time I retired, it had blossomed into a fully-fledged Naval Base.
Source credit: *Crown copyright*

1998. Retiring from the Navy, Kate at my side, after thirty-seven years service. Thanks to nuclear deterrence, I never had to fight a third world war.
Source: *Crown copyright*

Though the escalation continued, it was not mindless. It was the endless cost escalation that finally broke Soviet resolve. The nuclear arms race was devastating their crumbling economy. That was what ended the Cold War. The job of our Deterrent submarines was simply to maintain the nuclear stalemate until peace could be agreed. It took forty-five years.

Herein lies humanity's greatest conundrum: none of the nuclear powers had any wish to use nuclear weapons, but none would be without them. None would unilaterally disarm because, without nuclear deterrence, a third world war would be more probable. Such enormous nugatory expenditure seems utterly ridiculous but that is the price of peace. The horrors of the First World War had not deterred non-nuclear man from descending into the even greater horrors of the Second World War, a mere twenty-one years later. The cost of war is paid in blood and the price of peace is deterrence. As the Ancient Romans observed: *si vis pacem para bellum – if you wish peace, prepare for war.*

The order to launch a British nuclear counter-strike is a political, not a military, decision. It lies exclusively with the Prime Minister. It is not even a Cabinet decision, let alone a parliamentary decision. There would be no prospect of either in the aftermath of a Soviet first strike as Westminster would be in ruins. Therefore, on taking office, every Prime Minister selects three Nuclear Deputies from his ministers. They are appointed to take over the firing decision should the Prime Minister be killed. (Given enough warning, the Government and its top military advisers could potentially decamp from London to improve their odds of survival).

If the Prime Minister were not killed, the firing order would be transmitted from an ultra-top-secret Nuclear Operations Centre in a nuclear-hardened bunker. From there it would be sent to CTF 345 at Fleet Headquarters in Northwood, the Polaris operational control centre, thence to the submarine(s) on patrol. This is of course a gross simplification of an immensely complex sequence of encrypted messaging procedures. The time taken for the Prime Minister's order to reach the submarine(s) could be up to forty minutes. The submarine(s) would fire fifteen minutes later and the missiles would arrive on target some thirty minutes after that, an eighty-five minute time lag. Compare that with the four-minute warning of a pre-emptive attack.

Polaris submariners were therefore no more than the finger on the trigger of the long arm of a democratically elected government. But if London had been reduced to rubble, there was an implicit risk that the Prime Minister's firing order could not be sent. To deal with this possibility, every new Prime Minister, on assuming office, writes a personal sealed letter addressed to the SSBN Commanding Officers and it is carried on board the submarines.

In these letters, the Prime Minister gives instructions to the Commanding Officers on what to do if all normal communications are lost, 'lost' being taken as four hours with nothing heard. This is called the Letter of Last Resort, and sometimes as the Letter from the Grave. It is kept in a safe within a safe in the submarine Control

Room. One such letter was held in *Revenge*. It was kept inside a covering letter from the Commander-in-Chief, also in a sealed envelope, giving the Commanding Officer instructions on when to open the Prime Minister's letter.

As far as I know, no Prime Minister's Letter has ever been opened, nor have the contents of any ever been disclosed, but the possible options are obvious: retaliate, do not retaliate, seek refuge in a friendly country, or let the submarine decide. These letters are destroyed without being opened every time the Prime Minister changes. Only the Prime Ministers know what instructions they have given, and that offers scope for speculation.

Harold Wilson, for example, had been openly opposed to nuclear weapons – it was in his Labour Party manifesto – and some suspected him of having Soviet sympathies. He could not refuse to shoulder his Prime Ministerial responsibilities for national security but he could have selected 'Do not retaliate' in his Letter of Last Resort. In 2015, Jeremy Corbyn, a lifelong nuclear activist and one time Vice Chairman of the Campaign for Nuclear Disarmament, became leader of the Labour Party. Were he to be elected as Prime Minister, he could, hypothetically, select the 'Do not retaliate' option. If a potential aggressor were to be aware of this, our Independent Deterrent would have lost all credibility. From a national security point of view, the content of the Letter of Last Resort must never be revealed. Potential enemies must always believe that intolerable nuclear retaliation will be the inevitable consequence of their own first strike.

Did any of this bother me as I joined *Revenge*? Not in the slightest. I was only in a supporting role and not in the actual firing chain. Nevertheless, I would be complicit in any missile launch – but I felt no greater sense of guilt than anyone else in Britain who had voted for the Conservative, Labour or Liberal Democrat parties. All elected governments since the end of the Second World War have maintained our Independent Strategic Nuclear Deterrent and all our Prime Ministers have shouldered their nuclear responsibilities.

Much more importantly, I had complete faith that Deterrence worked. Mutually Assured Destruction is a thoroughly compelling principle; it is not a game of bluff. The Soviets were acutely aware that we were out there and had the ability to hit back. Such matters were never discussed on board but I believe that we all subscribed to this view. It was another nuclear conundrum. There we were in the most warlike posture possible, attack drills honed to a fine edge, standing ready night and day to launch Armageddon at fifteen minutes notice, but never believing we would have to do it.

Also, I viewed the Soviet Union as a responsible adversary. Their political machine was not like that of the Nazis. They calculated very carefully before they chanced their arm and knew very well that a nuclear Third World War would be disastrous for them. Setting aside their desire to proselytise Communist ideology, I believe that the Soviets were intrinsically defensive in their thinking. The Second World

War had hurt them far more than it had hurt Britain or America and they had far greater reason to avoid another.

So, how did mankind end up with such inhuman calculations? It was, I suggest, a combination of fear and evolutionary pressure: the atomic bombs dropped on Japan had shown the power of nuclear weapons, the iron curtain had descended, a nuclear arms race had followed, the advent of the hydrogen bomb and intercontinental ballistic missiles (ICBM) had led to universal vulnerability, and the arrival of ICBM-carrying nuclear submarines had removed the prospect of pre-emptive elimination. No one wanted to fight a nuclear war but no one wanted to drop out of the race. We were like intelligent lemmings, racing to the cliff edge but none jumping.

There were some attempts to fence off the cliff but with little success. Strategic Arms Limitation Talks (SALT 1), agreed back in 1969, limited Anti Ballistic Missile defences but failed to go any further. A Strategic Arms Reduction Treaty (START) was finally agreed between the United States and Russia in 2011, after the Cold War ended. Britain was not a signatory to that being already at minimum credible nuclear response level.

Throughout my life, I had been revelling in the joys of peace and freedom under my nuclear umbrella. Could it be, I wondered, that nuclear weapons were God's last great gift to mankind? Weapons so terrible that they could force us to live in peace out of fear? Were they man's last chance? What was the official Christian position? The Protestant churches were equivocal. The Catholic Church has not yet identified a saint for nuclear weapons. It was the Russian Orthodox Church that answered the question. When the atheistic Soviet regime collapsed, the re-instated Russian Orthodox Church declared Saint Seraphim of Sarov to be patron saint of Russian nuclear weapons 'both strategic and tactical,' the duality no doubt being an important ecclesiastical point. According to the Russian Orthodox Church, God does approve of nuclear weapons. Sarov, known as Arzamas-16 during the Cold War, is home to the All-Russian Scientific Research Institute of Experimental Physics. It is where Soviet nuclear weapons were developed.

****

In her awesome endeavour, *Revenge* was at the top of the country's Defence priorities. We had to be truly credible, to be ever ready to launch our missiles and so, if we wanted something, we got it. We even had a ton of sand delivered when our Stores Assistant entered a wrong number in the stores computer.

One of our other more unusual demands was for pickaxe handles. These were used to arm our below-deck sentries because, unlike Hollywood, we considered firing guns inside a submarine to be a bad idea. Not being trained as marksmen, our above-deck sentries were unarmed but a threat assessment proposed that they should be armed with sawn-off shotguns, their task being simply to drive invaders off the casing. I suggested reversion to blunderbusses but these were no longer stocked by Naval Stores.

The atmosphere on board was sober, focussed and ultra professional – an SSBN is no place for braggadocio. Commander Geoff Jaques, my new Captain, was extremely good-humoured and had a prodigious gift for restoring calm in a crisis. The Polaris Systems Officer, Mike Davies, had been at Manadon with me and was serious and cerebral. I had understudied him in *Onyx* during my submarine training and we got on well together. On the other hand, John Osborne, the Marine Engineer, was a volatile Irishman. He too had been at Manadon with me and knew me very well. He had also been in my submarine training class but had gone nuclear from the outset and was now a highly experienced nuclear engineer. He had also been on the staff at Dounreay during my nuclear training and had been privy to staff doubts about my performance in the simulator. I happened to be a few months senior to him in rank but was a mere beginner compared with him in the nuclear world. Herein lay a time bomb. When my predecessor left, I would become *de facto* Senior Engineer, thus inadvertently robbing John of the top job which made me feel like a gatecrasher.

If John was highly strung, John Hodgson, the officer I was relieving, was the complete opposite. He was a relaxed Cambridge graduate who took great pains to explain to me the Pareto Principle, namely that ten per cent effort gains a ninety per cent return whereas a hundred per cent result requires a further ninety per cent effort. It sounded like the theoretical basis for taking it easy.

The two Assistant Engineers were both impressive: Tim Chittenden was a direct entry Cambridge graduate who had worked with the Central Electricity Generating Board before joining the Navy and would go on to become an Engineer Admiral while the delightful Peter Sanderson, my deputy, was a fellow Manadonian. I was considerably senior to both but, again, inferior in nuclear experience. Bless them, they both accorded me the fullest of professional respect. Bob Lewis, the Missile Officer, was the only non-graduate amongst the Engineers. He had been promoted from the lower deck and was an unusual character. As Assistant Polaris Systems Officer, he did not keep watches and would sit in the corner of the Wardroom knitting. He also brought with him a two-month supply of unread *Daily Telegraphs* and delivered himself a fresh newspaper every morning.

Unlike other submarines, SSBNs carried a doctor. Ours was Rod Robertson, a charming, Glenalmond-educated Scot. His was a difficult position. We had a hundred and fifty fit young men on board and so he had little trade. His main practical responsibility was to choose the seventy movies we would take on patrol. His actual role was to deal with clinical emergencies because a Polaris submarine would not breach its patrol for a medical case. If the need arose, he would have to exercise his skill as a surgeon or declare the patient dead and store the cadaver in the deep freeze along with our food. He also had an undeclared role in checking that the Captain and other officers in the missile firing chain were of sound mind.

The Seaman Officers were an equally impressive team. Barry Coward, the XO, had been Commanding Officer of *Onyx* and was a member of MENSA (i.e. had an exceptionally high IQ). Jim Foster, the Navigator, I knew from torpedo trials. Jim, the Captain, and the XO were the only people on board who knew where we were whilst on patrol. David Southcott, the Sonar Officer, was superb with his hands and crafted beautiful wooden ship models. All three would go on to command Polaris submarines. Again unusually for a submarine, we also carried a Supply Officer. He was our top candidate for being of unsound mind. He had completed Royal Marine Commando training and insisted on wearing his green beret on every conceivable occasion.

I fitted in very well with everyone except for having a sensitive but polite relationship with John Osborne, the Marine Engineer.

Just after I joined, I happened to meet Chris Temple, a fellow Electrical Officer, on the jetty at Faslane. I was surprised to see him, as I knew he was standing-by one of the new boats being built at Barrow.

'What are you doing up here?' I asked.

'The Lord has called me.'

I was impressed. I only dealt with an Appointer. 'I thought you were standing by a new boat building in Barrow.'

'I was, but the Lord called me to *Courageous*.'

'I heard you were building a house down there.'

'I am.'

'Then what about Mrs Temple?'

'The Lord will find a solution for her.'

I didn't know whether to feel embarrassed or concerned. He had abrogated all responsibility for his life and his wife to God. His Faith had immunised him from all life's problems. How wonderful. I hadn't the heart to tell him that The Lord had been moved by my decision to opt out of *Courageous*.

Before we sailed, each member of the crew had to provide a twenty-four hour urine sample in a plastic flagon, the medical experts being concerned about the long-term effects of high carbon dioxide levels on our blood acidity. As I was going up to Glasgow on sample day, I put my flagon in an old leather briefcase and took it with me. When caught short in Sauchiehall Street, I sought the nearest public convenience only to find that all the cubicles were engaged and so had to stand against the urinal wall and pee into my briefcase. I felt like a pervert.

That, however, paled into insignificance compared with an incident at sea towards the end of the patrol when we were required to give a repeat twenty-four hour sample. I was sitting in the Wardroom talking to John Hodgson when John Osborne came in, pink with rage.

'Have you just peed into the bottle under our sink?' he demanded, glaring at John.

'Yes,' John H replied. 'I thought it was mine.'

'No, it bloody well wasn't. It was mine. You've just fucked up my urine sample!' It was as if his flagon had been violated.

Before deploying on patrol, we spent a week at sea on pre-patrol Index (Independent Exercises), a shakedown period in which we exercised every conceivable emergency drill from fire, flood and famine to missile launch and torpedo attack. This was essential because SSBNs had two crews to maximise the hull's time at sea. Therefore, each crew, when going back on-crew, had not been to sea for three months and was rusty. Also, twenty-five per cent of the crew would be new joiners as part of the on-going turnover of personnel. During this time, we carried sea-riders, who were Squadron staff ordered to verify our readiness. It was a hectic and nerve-wracking time, especially for the engineers who had to participate in all the drills whilst also setting up new equipment, testing what had been overhauled, and keeping watches. The Index period was followed by a few days of frantic repair work and then we deployed.

****

Going on patrol came as a blessed relief. When the Captain ordered: 'Officer of the Watch, come below and shut the upper lid,' we knew we were locked in for the next two months (or more) and would be utterly incommunicado. Polaris submarines on patrol never transmit. We had become, in every respect, a closed community. We would have to deal with all problems on our own. But first we had to slip out of the Clyde and into the Atlantic undetected and that was not as simple as it sounds.

The Malin Head AGI had to be avoided (if we went via the North Channel) and it would know from Soviet intelligence reports when we were due to leave. It would try to intercept us and, if successful, would pass the information to a Soviet SSN lurking off the West of Ireland. The Soviets were desperate to know where our Deterrent submarines patrolled and would have dearly liked to trail us. We played the same game up North but with considerably more success. They never would succeed with us.

But the Soviets were not the only ones an SSBN had to avoid. It was also a cardinal requirement that we were to remain undetected by all other submarines and warships. To achieve this, the ocean was divided into a grid and through careful waterspace management, interaction with friendly forces was avoided. This was managed by CTF311 and CTF 345, two secretive operational control groups in Fleet Headquarters at Northwood, the latter for Bombers, the former for Fighters. As they also collaborated with the other NATO navies, this meant that any submarine or warship detected in our patrol area could be assumed to be hostile.

Once into the Atlantic, we went on to Quick Readiness Alert (QRA). We were now carrying the baton of Deterrence. From now on, we would remain ultra quiet and keep out of harm's way. No one would know where we were – except for the flock of seagulls that had clutched on to our floating wire aerial and were water-skiing behind us in a perfectly straight line.

I love the way that animals take advantage of man's inventions. Sammy-the-Seal in Loch Long discovered that if he humped his way up the easy slope of a nuclear submarine's tail in harbour, he could bask in the heat from its Engine Room. And mussels loved to nest in our heat exchangers. There they were warm and safe but they blocked the tubes and caused machinery to overheat, a submarine form of sinusitis. This provoked a Technical Memorandum from Central Plant Control Authority describing the entire life cycle of the lesser-spotted Atlantic mussel and instructing us to keep our shore-side sea intakes shut in harbour during the mussel mating season. Apparently, they were joining us from the jetty piles to have nooky in our tubes. Think of them next time you order *moules marinières*.

Repulse once returned from patrol completely white. Subsequent analysis found that she was coated with a marine organism found in her patrol area. The identity of that organism had then to be declared secret.

****

In World War 2, the patrol aims of a submarine were: *Sink. Burn. Destroy.* For a Polaris submarine, the aims were: *Remain Undetected. Maintain Constant Communication. Be at Fifteen Minutes Notice to Fire.* These were our sacred principles and our patrols were meticulously analysed to ensure that these aims had been met.

Remaining undetected was our first priority. It was of fundamental importance that we remained invulnerable to pre-emptive attack. Maintaining constant communication was essential if we were to respond to a Prime Ministerial firing order. Readiness to fire at fifteen minutes notice was our commitment to obeying that order. We lived by these priorities night and day. We had no licence to make any decisions of our own about the missiles we carried. We were on a one-way communication ticket. We could receive and obey but not reply.

To ensure that we remained alert, we received Weapon System Readiness Tests without warning at random intervals. These were dummy firing orders with our response times being recorded. We had no way of knowing if these signals were the real thing until they were decoded, so they triggered immediate Action Stations Missile and that always got the adrenaline pumping.

All very well but I had my own more immediate patrol aims: *keep the water out, the oxygen up and the heads fully operational.* The two sets of aims were entirely compatible. If we failed in my aims, we could not undertake the others.

It was Napoleon who said: '*An army marches on its stomach.*' A submarine may not march on its lavatory pans but the principle still holds. With two weeks of our eight-week patrol remaining, the Supply Officer announced that we were running out of toilet roll. Within twenty-four hours, there was none. This was panic buying submarine style. We had a bog roll famine on our hands. The Executive Officer launched an immediate search of the boat, which revealed, *inter alia*, a dozen hidden behind the Captain's bunk. Other private hordes were found stashed away all over the boat. One secret cache almost caused a fire as it was hidden behind a heater. When it comes to toilet rolls, it's a case of every man for himself. However, there was no truth in the rumour that the engineers had a secret supply hidden in the Reactor Compartment. That was clearly a missed market opportunity because glow-in-the-dark toilet paper could have been a big hit. For the remainder of the patrol, toilet paper was rationed.

At the end of the patrol when we came off QRA, we released a torrent of signals including the crucial 'Reactor State Normal' plus an accumulated two-month shopping list and list of defects for repair during the inter-patrol period. I also took over my new departmental responsibilities but, before my predecessor left, the Captain informed me that John Osborne would be taking over as Senior Nuclear Engineer pending my gaining more nuclear experience. That made complete sense. What stuck in my craw was that it had been discussed behind my back and presented as a *fait accompli*. I had a right to be consulted before my seniority was overridden. Had I been asked, I would have agreed. It would have been plain stupid to object. It meant that I was now responsible to a theoretically more junior officer but what the hell, our patrol operations were far more important than my dignity. I would be damned glad simply not to fail in my duty.

****

As the subsequent off-crew period came to an end, the nuclear engineers undertook their statutory week of refresher training in the simulator. I was relaxed about that. I had done it in *Conqueror* and regarded it as no more than practice, but my performance triggered more behind-my-back discussions and John informed me that I would be paired with our most experienced Nuclear Chief Petty Officer at sea. This was somewhat humiliating but I was now perfecting the art of humility. For the next two months I would be sharing a tiny two-berth cabin with someone who doubted my competence and that made for a strained relationship.

When the Starboard crew brought *Revenge* back from patrol, we went on-crew and plunged straight into an intense period of maintenance. That I took in my stride. My only worry was doing the right thing at sea in an emergency. In the event, the patrol passed without incident.

During the following off-crew period, Commander Paul Hoddinott, our new Captain, joined. He had been four terms ahead of me at Dartmouth and was a thoroughly good hand. He was also red-haired and after my *Otter* and *Courageous* experiences with red-haired captains, that could have rung alarm bells. An exceptionally rigorous new Commander (E) had also taken over in the simulator. The first I knew of this was when I was sent for by the Squadron Engineer after my refresher session and told that I would be 'on probation' during my next patrol. I blew a fuse at that. This was becoming tiresome. It was two years since I had qualified as a reactor plant operator. I had spent six months in *Conqueror* and two patrols in *Revenge* without putting a foot wrong, yet I seemed to be trapped under the nuclear microscope. Nowhere in this interview was there a word of encouragement. The Squadron Engineer was merely covering his back so that if I did make a mistake, he could claim to have had me under special monitoring. I was not having it. If he didn't trust me, he should have the courage to remove me. And who was to be my probation officer? I could stomach temporary surrender of seniority but was not prepared to accept a subordinate as my probation officer. In any case, what action could a probation officer take on patrol?

This was the Rickover effect in action: no concessions were made to personal dignity; nuclear competence was the only criterion. As a result, some truly good guys were crucified. Two very close friends, both damn fine officers, had recently been dismissed from their submarines under this zero tolerance regime. One, hand-picked to be Senior Engineer of our latest boat in build, had been summarily sacked for not faring well in a verbal qualifying board and was replaced on the spot by his own deputy. The other was dismissed because a main generator bearing had wiped whilst in an American port. That was deemed to be a national embarrassment for which a head had to roll. In both cases, their careers were ruined. I seemed to be heading the same way.

I had wanted to specialise in underwater warfare rather than be a nuclear engine driver. I had been pressed-ganged into this and had worked my socks off to reach the required standard. I had even withdrawn my resignation to serve in *Revenge*, but the nuclear programme didn't do gratitude either. What I needed was encouragement, not constant undermining. Call me weak if you like but I flourish under encouragement. Most people do.

I went immediately to Paul Hoddinott to assure him that he could count on me to do my best but I could promise no more than that. 'If others consider me inadequate,' I said, 'I should be relieved of my duties. If not, I should be given moral support and not have the sword of Damocles hanging over me at sea.' Either way, I was not prepared to go on patrol on probation. 'It was,' I added, 'a simple case of back me or sack me.'

To save him a difficult decision, I handed him a letter of resignation explaining that I was perfectly prepared to continue but not on probation. I was merely giving him the bullet for my *coup de grace*. As I trusted him, I felt no need to have my letter logged in this time. He looked me in the eye and said: 'You're coming.' That was a morally brave decision for he was sticking his own neck out. If I fouled his pitch, his career would also suffer, but Paul understood the principles of leadership. I was a loyal officer and he knew it was right to back me.

He had never seen me in action as a nuclear engineer but had made a character judgment. I felt reassured.

Unknown to me, Paul referred the issue of military seniority versus nuclear experience to Submarine Headquarters, where it was ruled that the former took precedence. I should have been Senior Engineer. (This issue arose only during the rapid Cold War expansion of our nuclear submarine flotilla when some officers were parachuted late into the nuclear programme as I had been).

Just after we sailed, John Osborne turned to face me in the cabin. We were standing virtually nose-to-nose. 'I have a duty to monitor your watch-keeping,' he said, 'and to expose any incompetence.' So much for a perfect working relationship! He was my closest colleague and a brother Engineer. We had to work together. We were cabin mates. We shared a double-decker bunk. Now he would be my self-proclaimed probation officer. I had no idea whether this was his idea or if 'Big Brother' in the Squadron had tasked him with it. Perhaps he feared a mistake by me would rebound on him, or perhaps he just wanted to fire a warning shot across my bow to keep me on my toes. I don't know, but it was the sort of behaviour that the nuclear programme induced. Whatever John's reason, it was a mind-blowing statement. I took the hit on the chin and said nothing but it soured an already sensitive relationship. (Years later he told me that Big Brother had tasked him).

It felt as if I had been condemned to spend my entire career defending my competence. Like the Ancient Mariner, I had an albatross around my neck, but what on earth had I done to deserve it? I had always done my best, had never let anyone down, and had never actually failed. Now I had to face a two-month, white-knuckle trial of strength in which my antagonist was infinitely better armed. This latest tension was to be strictly man-to-man. John had been completely honest in telling me to my face that he was on my case and that could not have been easy for him. As we were both men of honour, the matter remained private. No one else knew, not even the Captain, but the remark put me under considerable extra pressure. I felt that everything I did was under his scrutiny and it doesn't half put a man off his stroke when he has to keep looking over his shoulder. Life was tense.

As we settled into the pedestrian routine of an SSBN on patrol, I began to sense that, despite his considerable experience, John was feeling the strain of nuclear responsibility. He was utterly competent but seemed averse to doing anything that might upset the plant. We did only the prescribed number of training drills like scramming the reactor and when we did, they were treated as major events that he presided over personally. He knew very well that a training drill could easily backfire if a trainee took the wrong action.

From a selfish point of view, this conservative approach meant that I was never really tested. Unlike in *Otter*, John would not be surreptitiously switching off power supplies in the middle of the night while I was on watch but on the downside, I was gaining little hands-on drill practice. Fortunately, in our slow motion world, things rarely went wrong – but one never knows when a real emergency might occur. One did.

I was in my bunk when the General Alarm sounded.

Baaaa Baaaa Baaaa. A voice rang out over Main Broadcast. 'Flooding in the Engine Room,'

Nothing strikes fear into the heart of a submariner more than the cry of 'flooding'. It means that water is coming in and threatening to sink the boat. Remember, a dribble is only called a 'leak.' I was off my bunk like a shot and heading aft. There was nothing I could do but instinct drove me. I charged through the Missile Compartment like an Olympic sprinter, bumping my shoulder against missile tubes on the way. The boat was already developing a stern down angle. I imagined the Engine Room filling with water. We were becoming tail heavy like the ill-fated *Thresher*. In reality, we were climbing. The Captain was taking us up to the relative safety of periscope depth.

At Checkpoint Charlie, I came to a halt. The bulkhead door was shut. Should I open it? That seemed like a damned stupid thing to do in a flooding situation. I looked through the peephole into the Reactor Compartment tunnel. The bulkhead door at the other end was also shut. This was an airlock so it would be safe to pass through. At the far end of the tunnel I peered through the peephole into the Manoeuvring Room passageway. There was no sign of life or of anything unusual. As I had shut the first bulkhead door behind me, I decided to go through.

When I entered the Manoeuvring Room, I was met with a row of ashen-faced watch-keepers, all focussing intently on their instrument panels. They had just been through a full flood emergency. A drain-cock on a seawater system in the Engine Room had blown and sent a jet of high-pressure seawater horizontally across the boat. John was on watch and had strangled it at birth by shutting the relevant hull valves. In turn, that had shut off cooling water to machinery, leading to a rapid sequence of machinery changeovers. The panic was over and the team were tidying up. It had been a minor but nevertheless genuine flood incident, more than enough to fire up the adrenaline system, and a salutary reminder that despite our pedestrian lifestyle, we were inside a steel tube and living in the depths of the ocean.

****

Every evening in the dogwatches (late afternoon), I called on the Captain to give him a report on the state of my department and anything else that I thought merited his attention. With my diesel boat experience, I understood his needs and he seemed to appreciate that. On one such call, he floored me by saying: 'I've decided that you will take over as Senior Engineer at the end of the patrol.' I was stunned. I didn't know whether to laugh or cry. From a vanity point of view, it was pleasing. From a practical point of view, it was terrifying. I didn't know how John would react. If I were to make a bad call, he could easily wield the dagger. To his enormous credit, he accepted the change without rancour and thereafter was entirely supportive. He never again questioned my competence and, to my credit, I made no mistakes. In the final analysis, we were both under pressure.

CHAPTER 16

Senior Engineer

'England expects that every man will do his duty.'

NELSON

Life in *Revenge* ranged from the ultra serious to the utterly ridiculous. During the next on-crew period when the Lord Lieutenant of Dunbartonshire was guest-of-honour for dinner on board, our gangways had been removed to speed up the submarine's delayed departure from the floating dock. Our guest had therefore to depart in a painter's bucket and be swung across to the jetty using a dockside crane.

The scenario turned to farce as the Captain and Officer of the Day stood to attention and solemnly saluted the bucket as it was hoisted into the air, the Quartermaster, in best Naval tradition, piping our guest over the side. From the bucket the Lord Lieutenant returned the salute and shouted down from on high: 'This is no way to leave one of Her Majesty's Ships.' The Deterrent programme was sacrosanct. It waited for no man. It could not be altered, not even for visits by the great and the good.

A few nights later, I was back aft in overalls doing rounds of the machinery spaces, hands covered in oil, when the General Alarm sounded followed by the cry: 'Lady in the Water.'

I had never heard that before. I had heard 'Man Overboard' often enough but never 'Woman Overboard' let alone 'Lady in the Water.' Nevertheless, the pipe had a ring of authenticity. So I rushed forward to investigate.

I found the Control Room packed solid with Chiefs and Petty Officers with their wives and sweethearts, all queuing to disembark after a party. As politely as possible, I clawed my way past them and started to climb the vertical ladder through the main access hatch only to find that I was climbing inside the skirt of a well-proportioned lady with pristine white knickers. She was having a panic attack and was stuck halfway through the hatch. As there seemed to be some sort of emergency taking place up top, I placed one oily hand in the centre of her pristine white knickers and up through the hatch she went sitting on my hand.

Having disposed of the blockage, I rushed out on to the upper deck to find the trot sentry tending the line to the ship's lifebelt, which had been thrown into the water. There, Petty Officer Robson and his wife were clinging to the lifebelt and having a domestic. When I asked the young sentry what had happened, he explained that Robson had been too impatient to wait in the queue below and had taken his wife up the conning tower and down inside the fin to gain quicker access to the upper deck through the fin door. He had however failed to warn his wife that the fin door opened only on to a narrow catwalk. So the unfortunate Mrs Robson had stepped boldly forth into the night and descended ten feet into the icy waters of the Gareloch.

Faced with the prospect of his wife drowning, Robson had screamed at the trot sentry: 'My wife's fallen overboard. Save her.'

'She's your wife,' replied the young sailor. 'You save her.' With that, he had pushed Robson into the water and thrown him the lifebelt. That was just before I arrived.

As it was not possible to haul the couple up the submarine's steep side, I gave instructions for them to be towed round to the bow and pulled back on board over the bull nose. When they had been rescued, I despatched them in the ship's Landrover to the Sick Bay in the Base where they were divested of their wet garments and given dry pyjamas to wear while an ambulance was arranged to take them home. But the somewhat agitated Mrs Robson would not wait for an ambulance and stomped off to the main gate to take a taxi. There she was arrested by the Ministry of Defence Police for wandering about the Naval Base in a nightie. When questioned, she said that she had been to a party in *Revenge*. The Vice Squad then came on board in search of an illegal pyjama party.

That was the end of the episode but the thing that intrigues me still is: how did the woman stuck in the hatch, explain to her husband my oily fingerprints on her pristine white knickers?

****

I found being Senior Engineer a damn sight more relaxing in practice. My imagination is too vivid. Without the actual responsibility, I was always conjuring up nightmare scenarios in which I would fail to do the right thing. Saddled with the real responsibility, I simply concentrated on the problem in hand. It came naturally to me. I even negotiated simulator training without criticism.

The submarine nuclear programme had evolved a personality cult centred on the Senior Engineer, evolved from a combination of the old submarine tradition of the Captain being the only star on stage plus the Rickover Effect. The concept was that, whatever the plant emergency, the Senior Engineer should take centre stage and deal with it. I had not agreed with the infallibility of the Captain principle, nor did I agree with it being applied to Senior Engineers. I could never credit myself with such supreme powers. It was obvious to me that the concept was flawed. The Senior Engineer may be asleep in his bunk when disaster strikes and watertight bulkheads could be slammed shut. He may not be able to get to the stage, let alone be the centre of it. And even if he could, he would be arriving in the middle of a dynamic event without knowing the preceding sequence of events. His intervention could be counter-productive.

I disregarded the Senior Engineer cult. I worked on the assumptions that I may not be present, that whoever was on watch must be able to deal with the emergency on his own, and that none of my Engineer Officers of the Watch should be less competent than I. With these three assumptions I knew where I stood.

I also expected that every man would do his duty, to quote Nelson. I needed to be sure that none of my watch-keepers would panic in an emergency and hit the wrong button. So, I introduced a comprehensive regime of training drills ranging from the minor to the major, in which every watch and every operator rehearsed all the emergency actions until they became second nature. 'Rehearsal' was the operative word in my lexicon. Training drills should be seen as practice, not testing. Take away the fear factor and most emergency drills involved only simple machinery changeovers, but in a panic errors can be made in simple procedures. If the wrong button is pressed, minor problems become major crises. I'd seen it happen.

In essence, the issue boiled down to a fundamental difference in leadership philosophy between that of being the star player and that of being the team manager. I subscribed to the latter, but also had to be a competent player. To my delight, my regime proved hugely popular. Watch-keepers would even demand extra drills to get ticks in all their boxes and I have no doubt that my team became one of the most proficient in the Navy. For my own part, I could sleep easy, safe in the knowledge that my team could cope without me.

At risk of sounding pretentious, this was the Nelson approach. Before Trafalgar, Nelson had his Captains row across to HMS *Victory* for a personal briefing. Thereafter, he had every confidence that they would do the right thing without further input from him, and he made his point in the most tragic way. He was mortally wounded at the beginning of the battle but his team went on to win the most famous victory in naval history.

****

Back at home, the BBC had been running a television series featuring the aircraft carrier *Ark Royal*. In researching for my own weekly radio programme, which included naval history, I learned that there had been an *Ark Royal* in the Royal Navy continuously since Elizabethan times. It set me thinking that in old naval dockyards like Devonport, there were probably families that had served the Navy continuously since the days of Drake. One night on watch, I asked the eight men gathered in the Manoeuvring Room how many had fathers who had been in the Navy. The answer was seven. The eighth had an uncle. That is one of the great strengths of the Royal Navy – sons, and now daughters, follow their fathers into it and that sense of duty is built-in.

Although restricted for sport, I also organised an Underwater Olympics, each department fielding a team. Events included arm wrestling, a sit-down tug-o'-war and an eating competition for which the chefs concocted the most disgustingly unpalatable grub. I watched men forcing the stuff down their gullets in a race to empty their plate, only to throw it all up again in one long sausage.

We also had the ever-popular inter-mess quizzes, a 'Brain of Revenge' competition, and a Sods' Opera, the latter causing me some concern because one of our youngest mechanics – I doubt if he had even started shaving – dressed up as a woman and looked absolutely gorgeous. It's frightening what eight weeks without female company can do to a man's judgement. For the rest of the patrol, pin-up photos of the young man began to appear all over the boat. Although it was all a bit of a laugh, I did worry about the risk of longer-term psychological damage to our young transvestite. But he was more streetwise than I. He had been a bookie's runner before joining the Navy and was older than he looked.

My creative writing skills also advanced. I included in my show an adult fairy tale about Julian the Jellyfish who was sucked into our cooling intakes and travelled through the ship's systems causing all sorts of mayhem. He became the explanation for every inexplicable malfunction. It was astonishing how our malfunctions could be matched to the journey of this fictional jelloid. I also created a White Rat letterbox into which scandalous betrayals of shipmates were furtively posted. After censoring, I would read them on 'The News'. A submarine is just like a village – we had gossip, spats and tit-for-tats.

Our Chief Radio Supervisor was a prima donna because he handled all the incoming signals. This meant he had read all our personal familygrams, and so was in the know about everything, which gave him an air of superiority. To cut him down to size, the Chief Stoker painted little gold stars on his leather sandals while he was asleep and thereafter he was known as Twinkle Toes. In retaliation, he delivered the Chief Stoker's personal familygram cut into individual words. In counter-retaliation, he found himself trapped in his bunk inside a curtain of high strength masking tape. In counter-counter-retaliation, the Chief Stoker's sleeping bag was then fed into the gash compactor and reduced to a solid cube. Such were the thrilling stories in my weekly News.

My News also included a weather forecast but not for the weather outside. What we needed to know were the oxygen and carbon dioxide levels inside. I was responsible for atmosphere control and had to keep us in an oxygen band between nineteen and twenty-one per cent, and at a carbon dioxide level of below one per cent. Outside these bands, we were into the death zones. At the lower end of the oxygen scale, the crew became grumpy and lethargic whilst at the upper end, they became high-spirited and energetic. Thus, if I wanted work done, I turned up the oxygen. If I needed peace and quiet, I let it drop. It was like playing God. Unlike in a diesel boat, we had no need to stick the snort mast up to suck down fresh air. We had electrolysers to split seawater into hydrogen and oxygen, and three large chemical scrubbers to remove carbon dioxide. These were our gills. As one wag remarked: 'When the canary shits gallstones, it's time to run another scrubber.'

On a more philosophical note, a nuclear-powered submarine could be regarded as a metaphor for our planet travelling through space, both having fixed and fragile atmospheres which are being continuously polluted. If mankind continues to pollute Earth's atmosphere and cause climate change, he may, one day, require massive nuclear power stations. This wouldn't be to generate electricity for man to squander but to drive massive carbon dioxide scrubbers and oxygen generators to restore the planet's atmosphere. Fossil fuelled power cannot do this as it produces the pollutants. Mankind cannot afford to ignore nuclear power.

****

There were only two untoward incidents on this patrol. The first was when our illustrious Supply Officer, he of the Green Beret, discovered that he was running out of food. This was crass incompetence as SSBNs had food lavished on them to maintain crew morale on our long and boring patrols. Two-thirds of the way through the patrol, he decided to reduce lunch to soup and a main course, which annoyed me because pudding was my favourite course. I said nothing until one day when I was last in for lunch. What was placed in front of me was a triangle of fried bread swimming in a puree of gravy with only limited evidence of mince. As a matter of principle, I never complain about food – some people are starving – but on this occasion I did. As the Supply Officer was sitting in the Wardroom, I commented in a loud voice: 'They had better food in the trenches.'

The remark lit his blue touch-paper – I must have given him too much oxygen. Immediately, he came over and faced up to me. I stood up and faced him, ready to defend myself if necessary though I had no Commando training. Before we came to blows, the Executive Officer was across and standing between us.

(Later, he explained that he had cut back on food as the patrol may have been extended).

That was only the second time in my long career that I had been close to unarmed combat. The first was in *Conqueror* when we were at anchor off Lossiemouth and all those who could be spared had gone on a run ashore. At the end of the evening, with much alcohol having been consumed, we were assembled on the jetty waiting for the liberty boat to take us back to the submarine when one of our young sailors tried to push me into the water. Fortunately, I caught sight of his drunken lunge from the corner of my eye and took avoiding action. I was livid and sorely tempted to thump him there and then but knew that it was neither the time nor place to make a scene. Besides, striking a junior rate would land me in very serious disciplinary trouble.

The following morning, I summoned the culprit to the Reactor Compartment tunnel where we faced each other alone. I was still fuming.

'Last night you tried to push me into the water,' I growled.

'Yes, Sir.'

'That's a serious offence.'

'Yes, Sir.'

'I'm not going to discipline you but if you ever try that trick again, I'll fucking knuckle you! Dismissed.'

When he had gone, I had a quiet chuckle. He was bigger than I and I had never knuckled anyone in my life. Who on earth did I think I was? Must have been watching too many movies.

The other crisis was when our Sewage Overboard Hull Valve became blocked. That was a real crisis. With one hundred and fifty active human bowels on board, limitless fresh water for showering, and loads of dirty dishwater from twenty-four hour galley activity, it would not be long before corporate constipation would cripple us. Why on earth had the Submarine Design Authority not thought of giving submarines two anal orifices – we had two of everything else? We were now about to become the first Deterrent submarine to breach its patrol simply because it could not empty its bowels.

In the finest tradition of the Navy, the situation was saved by the courage and self-sacrifice of one young officer. Tim Chittenden donned an immersion suit and breathing apparatus and went SCUBA diving in the Shit Tank to clear the blockage. He should have received an Honour for that, be made a Knight of the Porcelain Throne, or given the Order of the Bath. Something like that anyway. It came as no surprise when he later became an Admiral.

Food and sewage crises apart, this was a very enjoyable patrol.

CHAPTER 17

My Last Patrol

'Do not take revenge, my dear friends, but leave room for God's wrath.'

One evening during the subsequent maintenance period, I went to the Captain's cabin to give him my daily progress report. As he had gone ashore, I left a note on his desk and noticed the seating plan for his dinner party that evening. At the head of the table was a Commander Kitcat, an uncommon name that stirred childhood memories. So I scribbled that my father had served under a Commander Kitcat during the war, adding it could have been either in HMS *Wanderer, Seychelles,* or *Intrepid.* 'Would your guest be related?'

That evening the Captain phoned me at home. Not only was Commander Kitcat related, he *was* my father's Captain when *Intrepid* was sunk at Leros and was most keen to meet one of his surviving crew. A re-union with my father was then hastily arranged. Kitcat had been too badly injured in the bombing to escape with his crew and now, thirty-five years later, my father was able to tell him of the crew's clandestine journey to Alexandria. In return, Kitcat was able to tell my father about the aim of their mission. It was an astonishing re-union. My father was then sixty-eight and Kitcat was older.

The coincidence became even greater. The reason for Commander Kitcat being my Captain's guest was that his wife was my Captain's godmother through Paul Hoddinott's father having been Kitcat's Senior Engineer. History was repeating itself.

By now, John Osborne had been re-appointed to the staff of Submarine HQ and I moved from one Celtic connection to another but of an entirely different sort. His successor was Paul Thomas, a brilliant, dynamic, ebullient, English Public School educated, self-confident Welshman who was excellent company. He had been two years behind me at Manadon and was now both an experienced and hugely enthusiastic nuclear engineer. (He would go on to head the Navy's nuclear design programme as an Admiral).

Life was now fun. We were both John Cleese fans and always entered and left our cabin doing a Cleese silly walk. Off-watch, Paul constructed an elaborate model railway system – I think I slept in Clapham Junction – whilst I wrote scripts for my weekly radio show.

A surprise visitor before we sailed was a lady called Margaret Thatcher MP. Astonishingly, she was the new leader of the Tory Party. 'Astonishingly' because she was a woman, State educated, a grocer's daughter and a chemist by profession. She could hardly have been more at odds with the Tory tradition of patrician, male, Public School educated leaders. Her reason for visiting was to be briefed on Strategic Nuclear Deterrence in her capacity as Leader of the Opposition.

As an avid spectator on the political scene, I was well aware of the lady's somewhat hectoring style and of her lukewarm support for State education in the post Grammar School era. When I found myself talking to her in the confines of a crowded submarine wardroom – I was close enough to have put my arms around her – all my pre-conceptions were blown away. She appeared much older than her television image and was not in the least hectoring. She was utterly charming, firmly switched to receive, and had an air of serenity. She seemed more like my mother than a future Prime Minister. My immediate reaction was: what is a nice, little, old lady like you doing in politics? I would soon find out.

Britain was the sick man of Europe, an economic basket case riddled with strikes and suffering from high inflation (28% in 1975), which seriously eroded my father's pension. The Labour Government was wrestling with the trade union leaders to control inflation through a national pay policy but the unions seemed to be calling all the shots. They had eleven million members in those days and thought they had the power to hold the country to ransom. We seemed to be a country in terminal decline and I failed to see how the situation could be reversed. I looked at the lady who sought to be our next Prime Minister and asked her the simple question: 'How could you ever defeat the power of the trade unions?'

She looked me straight in the eye as if talking to a friend at a local coffee morning and replied with utter conviction: 'If the unions want confrontation, by God I'll confront them,' a phrase she repeated later in a BBC interview.

Fine words, madam, I thought, but how on earth will you do that?

****

As this was to be my last submarine patrol, I decided to grow a beard, simply to give the boys a surprise when I returned home, conscious that the last time I had grown one was back in *Otter* and that had been a career disaster. Some may say I was tempting Providence.

Providence, it seemed, had indeed been provoked. By our second week on patrol, we were having regular Reactor Compartment bilge high-level alarms. As there were no seawater connections in there, the bilges should have been dry. It meant that there was a steam leak. But was it radioactive steam leaking from the reactor or ordinary steam leaking from a boiler?

The issue had echoes of the Soviet *K19* disaster, re-named *Widowmaker* in the post Cold War Hollywood movie but nicknamed *Hiroshima* by her crew. On her first patrol off the East Coast of the United States, *K19* suffered a serious reactor coolant leak and to prevent reactor meltdown and save shipmates from certain death, seven volunteer crewmembers worked in a lethal radiation environment to repair the leak. Within days, all seven had died horrific deaths from radiation sickness and another twenty would die later. To what extent this courageous repair team was self-sacrificing and to what extent merely ignorant of the consequences of working in a radioactive environment is open to question. Certainly, they were led incorrectly to believe that their chemical protection suits would protect them. That was patently untrue. Nothing could have saved them. It had been like cooking them in a microwave.

More of *K19*'s crew would have been saved if an offer of American help had been accepted but her Captain preferred saving Soviet face to the lives of his men. There was no lack of courage in the Soviet navy but there was a criminal lack of duty-of-care by the Communist regime.

Steam and heat can be sensed instantly but radiation is a silent, odourless, invisible, insidious, whole-body-penetrating killer. One can be in a lethal radiation environment and initially be completely unaware of it. By the time one finds out, the damage is done, sunburn being a simple example. In our case, entering the Reactor Compartment required not only a preliminary survey to confirm tolerable radiation levels but also the wearing of personal dosimeters through which the dose being received could be read and kept within tolerable limits.

Being a firm believer in the stitch-in-time principle, I decided that prompt action was required and Paul Thomas agreed. We were only ten days into an extended ten-week patrol and high-pressure steam cuts through metal like wire through butter. Without intervention, things would only get worse. Thus began a marathon repair job of a kind that had never before been undertaken in a British nuclear submarine on patrol. Given the excellence of our reactor design and the fact that no radiation alarms had sounded, I suspected a conventional steam leak but nevertheless, the problem lay within the radiation environment of the Reactor Compartment. Finding it may not be easy but fixing it would be even more difficult, perhaps not even possible without breaching the patrol.

First we would have to shut down the reactor, go on to battery power, wait twenty minutes for radiation levels to subside, and then make a Reactor Compartment entry. Battery capacity allowed a one-hour shutdown which meant we had only twenty minutes to search due to needing twenty minutes to take the reactor critical again. In the event, no search was required. As soon as the Reactor Compartment plug was opened, we were met with a wall of steam. Just inside the plug door, a gasket had blown on the Starboard Main Steam Stop. Thanks to Admiral Rickover's uncompromising demands for the highest of engineering quality standards, our reactors did not leak. Thanks to Ministry of Defence economy measures, our secondary steam systems did. (We used gaskets, not welded joints). The leak was far too serious to ignore. It had to be repaired.

It took two days to decide on a plan of action. With only twenty minutes working time available per Reactor Compartment entry, it would require seven or eight reactor shut downs to complete the task. But we could not even begin until the starboard steam generator had been emptied, something which had never before been undertaken at sea and we would have to do this submerged on Deterrent patrol whilst maintaining patrol aims. It was an utterly abnormal operation achieved by draining the boiler down through a tiny sampling line. It took seven hours to empty, exactly as calculated.

Then began a surreal sequence of reactor shut downs, compartment entries, and reactor re-starts. I found it thoroughly bizarre to go into the tunnel on dived patrol and find the great, lead-lined plug of the Reactor Compartment wide open, as if we were back in harbour in mid-maintenance. Apart from the mechanical work, we also managed the radiation exposure of the repair team. Unlike the Soviet navy, I would not be sacrificing my men. Fortunately, as we had not been running at high power, radiation levels were low.

When the repair was complete, we came to another major hurdle: how to refill the empty boiler? This had never before been done at sea either and could only be done with hot water and the only way that could be achieved was by interconnecting the working boiler with the empty one and letting it supply the boiling water. The worry was that the working boiler was at lower temperature than the empty one. Were we trying to pour water uphill? Would the transferred hot water simply flash into steam, create a bubble and stop any further flow? If so, we would have no idea of where the water level was in the empty boiler and without knowing that, we could not use it. Paul and I attempted a thermodynamic calculation but were not convinced by the result. We would just have to suck it and see.

While the hot water was being transferred, we crouched side-by-side at the tunnel window and watched with bated breath for a level to appear in the boiler gauge glass. For what seemed like an eternity, the gauge glass remained empty. After twenty minutes of nothing, our hearts sank. We had failed. That meant we would be on a single boiler for the next nine weeks. That would be like going on a long route march with only one leg. Then, Glory Hallelujah, a level suddenly popped into the gauge glass as if from nowhere. We roared in triumph, hugged each other and punched the air. We had two perfectly good boilers again, an outstanding engineering triumph.

The procedure had taken thirty-six hours of continuous nuclear plant operations all set against the on-going operation of the submarine. I had hardly slept. At one point, I fell asleep sitting on a stool and had a nightmare. In this, I was standing in the tunnel confessing to the Captain that we had failed and had just become the first British Deterrent submarine to breach its patrol. Then I woke up and was completely disorientated. For a few moments, I thought the dream was real and had to figure out what was actually happening.

We had just achieved something no other nuclear submarine engineers had even attempted, at least not in the Royal Navy. We had successfully conducted multiple extensive reactor plant operations, had taken calculated risks, had ventured into the unknown, but had succeeded. In celebration, the other officers held a mess dinner in our honour.

****

The following morning, I had just taken over the watch from Tim Chittenden when there was a sudden roar. It sounded like a jumbo jet taking off – but there are no jumbo jets in submarines.

'Steam leak in the TG room!' a voice shrieked over the intercom.

The roar said it all. This was a major emergency. My moment of truth had come… *(see Chapter1)*

****

The consequence of this major steam leak was that we were left with only half our power generation system but because of complex nuclear safety rules, we were actually restricted to quarter-power. We could limp along at patrol speed but little more. After two tense weeks, it became clear that we needed more electrical power. We had to reclaim the starboard turbo-generator. It was technically possible to re-supply it but that would require fully scramming the reactor, cross-connecting the steam ranges, warming through the now cold starboard steam range and then pulling rods (restarting the reactor) with the port boiler feeding both steam ranges. Warming through a cold steam range normally took four hours. We did not have that time so we would have to do it in two, which was the absolute limit of our battery. It was a considerable risk.

Ruthless reduction in power consumption was invoked: lights were switched-off, crew sent to their bunks, air conditioning stopped, galley shut down, bilges pumped, the area checked clear of shipping, and our speed reduced to a mere creep. For two seemingly interminable hours we sat in eerie silence, suspended at depth with all machinery stopped, footsteps resounding from one end of the boat to the other like drum beats. This was the still of the deep. After two hours, I ordered reactor start-up. The risk had paid off. We still had only one boiler but now we had two electrical generators, a very welcome improvement.

****

Revenge was a happy ship. There were one-hundred-and-fifty men on board and we worked brilliantly well as a team. Paul and I were perfect cabin mates but not soul mates. How I longed to talk to Kate but she was weeks away. I felt an overwhelming desire to communicate with her. In desperation, I wrote this letter. It could not be posted but as I wrote, it felt as if I were talking to her. Was this how it would be when death did us part, in each other's thoughts but eternally unable to communicate?

Day 48

3rd June 1978

Sweetheart,
This patrol has been hellish. I find it hard to believe we still have three weeks to go. Five weeks ago, I faced the prospect of imminent death when a steam pipe burst, but my greatest fear was that I might never see you again. Even now I'm counting no chickens. So much has gone wrong. It's as if we've been cursed.

I've been living on survival levels of sleep and on occasion have feared a nervous breakdown through sheer exhaustion. My normal ten-and-a-half stones have withered away to nine-and-a-half. At the moment, we're limping home in a crippled state and that only as a result of superhuman efforts by my department. One more major fault and our limp will reduce to a crawl. We may even have to be towed home.

I've always told you that you are the most important thing in my life. Sometimes I've felt you didn't believe me, but when the chips were down, believe me, seeing you again was my only thought.

In the past, I know I have upset you. I feel ashamed of that and never want to make you unhappy again. I just want to love you and have you love me. I always say my prayers at sea. I don't know if God is listening but I pray for health, happiness and peace. Above all, I pray that I shall be a good husband and father.

It is so easy to be complacent when life is simple; so easy to take each other for granted. This should be my last patrol. If I survive it, we must resolve never to take our love for granted in the years ahead. Keep this letter and show it to me if you ever think I'm behaving badly.
With all my heart,
Eric

****

Operating a nuclear submarine is serious stuff but under normal circumstances, there is time for fun and games. Inter-mess quizzes are particularly popular, as are cribbage, liar dice, uckers (a submarine version of ludo), disc jockeying on the piped sound system, and watching movies. More intellectual members also play bridge or read. One SSBN

Commanding Officer even took a sports jacket and flannels on patrol and every Sunday after church, would put them on and go for a picnic in some obscure corner of the boat.

The Wardroom staged two formal mess dinners per patrol and, somewhat incongruously, it was on this patrol that I gave my first ever after dinner speech. It was for Trafalgar Night and I had to toast 'The Immortal Memory.' As I struggled to keep *Revenge* on task, the leadership and heroism of the great Admiral Nelson was firmly on my mind.

Surprisingly, there were great similarities between Nelson's navy and my own. His ships were powered by the wind and, like nuclear submarines, had no need to refuel. His endurance at sea was limited only by the food he carried, as with us. Nelson had no radio and could not communicate with the Admiralty back in London, just as we maintained radio silence. His task was to deter Napoleon from invading Britain by blockading the French navy in its own ports and threatening destruction if it ventured forth. Ours was to deter the Soviets from invading Western Europe by threatening them with nuclear destruction if they tried.

No one in Britain had known that Nelson was fighting the greatest naval battle in history. It was fourteen days before HMS *Pickle*, a fast topsail schooner, brought home the news of his great victory. No one knew of our struggles either. They would find out only when we came off Quick Readiness Alert at the end of our patrol.

Nelson, however, had to face the raging of the sea while we cruised silently beneath it. Also unlike us, he had to face the violence of the enemy, whereas my clash with the ruptured steam pipe had been the closest I had come to death. Nelson succeeded in his task. He saved Britain from Napoleonic invasion – but it cost him his life.

****

As one of my vices is practical joking, I had brought some exploding cigar tips on this patrol for insertion in the Wardroom Panatelas that were passed round after mess dinners. One night, I snuck the Wardroom cigar box into my cabin, removed two Panatelas from their tubes, unwrapped their cellophane, inserted the explosive tips, and returned the box to its cupboard.

The trick worked to perfection. At the Trafalgar Night dinner, there was a small explosion as the end of the Executive Officer's cigar blew-off in mid-puff, leaving him sucking on a tattered stump. The other spiked cigar was nowhere to be seen. In planning this joke, I had not considered that at the end of our patrol, we would be handing the boat over to the other crew, including the cigar box.

A tradition of the Deterrent programme is that a VIP meets every returning SSBN in the Clyde estuary and rides it back to Faslane. The VIPs could range from the Prime Minister down to senior Admirals. Another tradition is that after lunch, the VIP is invited into the Captain's tiny cabin for coffee and a cigar. Four months later, the Captain of the other crew was entertaining his VIP guest, the Commander-in-Chief, in the privacy of his cabin when the end of the great man's cigar exploded. Until writing this book, the perpetrator of that joke has never been identified. In military speak, it's called 'third party targeting.'

****

After ten weeks, we came off Quick Readiness Alert and re-entered the Clyde. *Resolution* was now out on patrol. Invisibly, we had handed over the baton of Nuclear Deterrence. My great chum, Henry Buchanan, was now fulfilling my role. We had gone through training together and he was one of Richard's godfathers. The Submarine Service is truly a close-knit community.

Coming off QRA meant that we could now signal to Submarine Headquarters and let them know that we had been having problems and were returning in an abnormal reactor state. Immediately, the electronic interrogation began. Had we done this? Had we done that? Headquarters was in a flap. They had to be sure we were safe to enter harbour.

On the eve of our return, I wrote again to Kate.

Day 69

15th June 1978

> *My Darling Kate,*
> *This should be my last night at sea in the Royal Navy. As ever, I'm thinking of you. I'm not feeling sentimental. I'm too worn-out for that. I've had little sleep. After ten weeks of hell, I'm completely numb, but want to pen a few thoughts while I have them in mind.*
> *In all my time at sea, I've never stopped thinking about you. Your constancy has given my tortured mind a permanent focus. I fear that I've made you unhappy by going to sea. Throughout our entire married life, you've had to suffer all the penalties of my chosen career. After this patrol, I shall feel content that I have done my duty to the Navy and the country. Then it's your turn. In the Penguin Dictionary of Quotations you gave me, I stumbled across one that describes exactly how I would like us to be:*
> *'Two souls with but a single thought,*
> *Two hearts that beat as one.' (John Keats)*
> *Let us be like that and not squander our love on superficialities, like two cabbages on a greengrocer's counter.*
> *I think we should discipline ourselves to have at least half-an-hour of dedicated communion every day. Even if we don't speak, at least we shall know that the appointed half-hour is devoted to thinking about each other.*
> *This patrol has reduced me to a zombie. I react mechanically to bells, buzzers and alarms. My emotions are dead. I can scarcely imagine another life. I can't believe I shall be seeing you tomorrow. I can't believe you are real. For the last ten weeks, you've been a*

figment of my imagination. Tomorrow I shall touch you again. I'm trying to remember what you feel like and how you will sound.
    All my love,
    Eric

****

On return from patrol, our first port of call was Coulport for the offload of selected missiles. Shortly before we arrived, the Captain released the bad news familygrams he had been withholding. Two were given to me for onward delivery and I then had the unpleasant duty of informing one of my best Chief Petty Officers that his wife had left him and another that his father had died. That was bad enough but Commander Bob Seaward in *Repulse* had the hellish task of deciding on patrol to withhold the tragic news that the wife and child of one of his crew had been killed in a car crash. The night before *Repulse* returned, he had to call the man to his cabin and reveal the awful truth. Can you imagine the agony of that news? After more than ten weeks on patrol, that poor man would have been overjoyed at the prospect of a reunion with his wife and child the following day, only to be told that they had been killed seven weeks earlier.

Once alongside, I clambered up on to the after casing via the engine room hatch for a lungful of fresh Scottish air. After ten weeks of breathing artificially generated air laced with lubricating oil, steam, cooking and sewage; fresh air was a heady cocktail of salt water, seaweed, and pine forest. The inside of the fin stank like a fish market.

Alongside our seaward side sat one of the tugs that had helped with our berthing. On its winch-deck stood a couple of young women wrapped up against the stiff southerly breeze. I could scarcely believe my eyes. One was Kate – how indescribably wonderful. How on earth had she managed that? She had never managed it before and it was pure chance that I had popped up on to the upper deck. She must simply have wanted to sight *Revenge* returning, knowing that I was on board. We were close enough for a shouted exchange of endearments but it was a case of so near and yet so far. She could not come on board and I had to go back down below to shut down the reactor, but I could not have wished for a better welcome home. She was shocked at how thin I'd become. Later she told me she had sensed for weeks that something was wrong.

The following morning we moved round to Faslane to be greeted by Squadron staff, but they were more like the Spanish Inquisition than a welcoming committee. The inquiry began immediately. Had I been at fault? Had I broken any nuclear rules? I didn't care. I'd lost 10% of my body-weight. I'd grown a beard. I'd done what seemed right at the time. We were home safe. The reactor and missiles were unharmed. Our mission had been accomplished. We had not dropped the baton. I was damned proud of our achievement.

We had not expected to be greeted as heroes. Our business was far too secretive for that. We had, however, expected full support from Base staff. What we returned to was strike action by the Civil Service industrial trade unions, led by the Transport and General Workers Union. Incredibly, these fellow Government employees were attempting to sabotage the Deterrent programme in support of a pay claim! Their intention was to stop *Revenge* sailing for her next patrol. Had they succeeded, it would have destroyed the credibility of our Deterrent. It would have shown the world that Britain's top strategic Defence priority was a hostage to the trade unions – exactly what the Soviets would want.

The unions were targeting the Deterrent to blackmail the Labour Government into abandoning its pay policy. To me, this was nothing short of treason, and to rub salt in the wound, officers and men of the Royal Navy, including those in *Revenge* who had just returned from patrol, had their leave cancelled and were drafted in to undertake the work of striking civilians.

These striking Civil Servants were taking the Queen's shilling just like the men in *Revenge* but without risking their lives or being separated from their loved ones. As members of the Armed Forces, we were duty-bound to serve our country. The strikers didn't even show loyalty to their own Labour Government, the Party of Organised Labour. Ever since that strike, the Navy has maintained a uniformed presence in all of the civilian departments involved in the Deterrent.

We didn't know it then but the country was about to enter the notorious Winter of Discontent (1978/79) when even the gravediggers would go on strike. That toppled the Callaghan Government. In March 1979, Margaret Thatcher called in Parliament for a vote of No Confidence in the Government and won. In the subsequent General Election, under the slogan of *Labour isn't working*, the Conservatives swept back into power with a massive majority. The people had spoken. With overwhelming public support, Margaret Thatcher rapidly introduced legislation to curb the powers of the trade unions. The union leaders had shot themselves in the foot. She would be Prime Minister for the next eleven years and it would be another eighteen before Labour returned to power.

Margaret Thatcher's Government, without hesitation, ordered the replacement of Polaris with the new American Trident 2 missile system. The outgoing Labour Government had been dragging its heels on this decision for years but, in fairness to Prime Minister Callaghan, he had secretly ascertained with US President Jimmy Carter that America would be prepared to sell. By an amazing coincidence, President Carter was not only an ex-submariner but had worked under the legendary Admiral Rickover in development of the US Navy's first nuclear submarines. (Jim Callaghan had been a Lieutenant in the Royal Navy during the war).

The trade union barons of the time were intoxicated with power. They thought they could hold Britain to ransom. They believed in socialist dictatorship through the bullying of the Labour government. They played on class prejudice and deluded their membership into believing that nationalised industries could pay unaffordable wages and run at a loss. They never understood that the democratic majority of British voters had no desire to be ruled by them.

I am not against trade unions. They evolved for very sound reasons and many honourable men and women have been trade unionists. However, in 1978, control of the unions had fallen into the hands of left-wingers and outright Communists who fancied Soviet style Socialism whilst seeking to milk the proceeds of our free Capitalist society. In my book, they were parasites. The Labour Party agreed. After eighteen years in Opposition, it had to purge itself of such extremists to make the Party re-electable as New Labour under the leadership of Tony Blair.

****

After interrogation by the nuclear safety gurus, I went home. During those seemingly interminable four-hour watches on patrol, I had often asked myself why I was not doing something more useful, like being a doctor? My answer was always the same. If, in my very small way, I were helping to prevent World War Three, then I would be saving far more lives than any doctor. I felt at ease with the thought that my life was not being spent in vain. I firmly believed that Deterrence was the best policy for peace. Blind hope never prevented war.

In the weeks following, *Revenge* was showered with praise as the nuclear experts digested our multiple Incident Reports. They could not fault us. In the closed world of nuclear submarine engineering at least, we were now hailed as heroes, and Emergency Operating Procedures were re-written to reflect our experience. But I was not fooled. Had things gone wrong despite our best endeavours, the authorities would have been calling for my head on a plate. I was particularly moved to receive a personal phone call from John Osborne to congratulate me. It was a truly noble gesture, the mark of a true officer and gentleman. He had read our Incident Reports in Headquarters.

In the New Year's Honours List of 1979, 'Bungy Mack', the young Leading Mechanic who had been on watch down below when the steam leak occurred and who injured his back in attempting to stop it, was awarded the Queen's Gallantry Medal. Lieutenant Commander Paul Thomas (later Admiral) and three of our merry men (Mechanicians McCrum, McDonagh and Murdoch) were awarded Oak Leaves for gallant conduct, and I was honoured with an MBE for 'technical leadership'. I could not have asked for more. My leadership and technical competence had been endorsed at the highest level.

The following February, Bungy Mack and I went to Buckingham Palace to receive our awards. When he arrived in the waiting area, ultra smart in his sailor suit, I could smell him before I could see him. He was reeking of beer. This brave young man had been so nervous about the investiture that he had resorted to Dutch courage. He proceeded to breathe alcohol over Her Majesty Queen Elizabeth, the late Queen Mother.

A young, six-foot-tall, Liverpudlian sailor breathing beer over Her Majesty clearly created a big impression, for She had a significantly longer conversation with him than with any of the other much more senior Honour recipients. Her Majesty even recalled the encounter later in an interview for a woman's magazine.

Acting Leading Mechanic James 'Bungy' McWilliams QGM was then twenty and I was a thirty-five-year-old Lieutenant Commander, almost old enough to be his father. He was a perfect example of how the Royal Navy achieves the very best from its young men (and now women). Bungy was a jack-the-lad from Liverpool, a mad keen football fan and wild as any young man of his age when ashore, but clap him in a sailor suit, put him through Naval training, stick him in a nuclear-powered submarine in an emergency situation, and he will commit his life to saving his ship. To say that young folk are not as good as they used to be is patent nonsense. The young are just as good as young people ever were. How they behave depends on how they are educated and the Navy gets that absolutely right.

As we left Buckingham Palace together, I felt as if I had just won my nuclear spurs. I had not failed when the chips were down. The albatross was off my neck.

CHAPTER 18

War and Peace

'They all seemed so able to demonstrate extraordinary qualities.'

MARGARET THATCHER, AS PRIME MINISTER,
REFERRING TO THE MEN OF THE SUBMARINE SERVICE

At sea, I had managed man and machine, the former a pleasure, the latter a constant anxiety. With my seagoing days now behind me, I felt a huge sense of release. I was unchained, ready to spread my wings. My sons were in the Cubs and when I watched them promising to do their duty, just as I had done thirty years before, I felt it was time to plough something back and volunteered to become a Scout Leader – and what rich rewards that brought. Teaching teenage boys to navigate through the mountains and forests of Argyll with a map and compass, to operate a motor boat borrowed from the Base round the lochs and islands of the Clyde, and mucking in with them at summer camp or at international Scout camps, were all thoroughly life enriching experiences.

In this new endeavour, I was united with Dr. Geoff Riddington, someone of my own age but from an entirely different background. He was a university Economics lecturer, a left-wing member of the Labour Party, anti-military, and strongly opposed to nuclear weapons. It was an interesting departure from the one-sided naval view of the world. We disagreed on virtually everything except our belief in the principles of Scouting. Nevertheless, we formed an extremely effective pairing and became great friends.

Many years later, when canvassing for a General Election, I was accosted by one of our former Scouts in the street. He was by then in his late twenties. 'I'm so grateful to you and Geoff for everything you did for me,' he said, 'but you must have hated each other. You never stopped arguing – but that was great for us because we all knew that if you said 'no,' we could go to Geoff and he would say 'yes'.' To quote Robert Burns: *Would some pow'r the giftie gie us, to see ourselves as others see us.*

With considerable academic trepidation, I also began a two-year part-time Masters degree in Acoustics at Heriot-Watt University in Edinburgh. As underwater warfare was almost entirely about acoustics, I would truly have upped my game if I could graduate in that. As the Navy refused to pay, I funded it myself and that concentrated the mind. I succeeded without difficulty.

In career terms, I was going nowhere, the *Courageous* incident being an indelible black mark on my record, and so I spent the next two years on the staff of the Tenth Squadron. Whilst there, I saw to my great pleasure the first four Tactical Weapons Engineers joining the Squadron straight from training. They were neither nuclear propulsion nor strategic missile engineers but underwater warfare specialists, exactly what I had called for back in *Osiris*. Whilst this was very satisfying, I still had some unfinished business in the underwater warfare world.

Our Deterrent submarines were the nation's top Defence priority but were not the only game in town. SSNs were playing a much more dangerous and comparably secretive Cold War role. Unlike SSBNs whose priority was to keep out of everyone's way, the role of SSNs was to get up close and personal with the Soviet navy. Since the Berlin Airlift in 1957, the Cold War on land and air had remained largely stagnant in the European theatre, but underwater it had been continuously hotting up. The underwater space had become the new front line. The US Navy's advances in nuclear submarines, particularly SSBNs, had put the Soviet navy in a very inferior position and throughout the Cold War, the Soviets were pouring massive resources into catching up. They sought superiority or at least parity with the West at sea and were building up a substantial flotilla of nuclear submarines in their Northern Fleet. They were based in the closed town of Severomorsk in Murmansk on the shores of the Barents Sea, an even more extreme location than Faslane.

The Soviets were producing design after design of submarine and weapon system in an effort to perfect their product. We needed to know how they were progressing. So our SSNs, working hand-in-glove with American SSNs, were tasked with maintaining continuous surveillance. In this endeavour, there were three main requirements. The first was intelligence gathering. Here, SSNs would sneak covertly into the Soviet's back yard and hoover up everything they could find about what the Soviets were doing. We intercepted their communications traffic and radar characteristics. Details of ships, submarines and weapon firings were observed and the acoustic signatures of Soviet ships and submarines were recorded during close encounters. For this work, 'special fit' boats carried additional data recording and analysis equipment along with specialist Intelligence personnel e.g. Russian speakers. The payback was that all NATO ships, submarines and maritime patrol aircraft were equipped with encyclopaedias of information on the electronic and acoustic signatures of Soviet ships and submarines, and on their tactics and weapon capabilities.

The most dangerous operation of all was doing photographic under-runs on Soviet ships and submarines whilst they were moving. This required the SSN to manoeuvre to within a few feet of the target's keel, a challenging operation that led to *Warspite*, *Sceptre* and at least three American SSNs suffering serious collisions. When two 'blind' vessels, each weighing around 5,000 tons, crash into each other, frightening levels of energy are involved; balletic *pas de deux* it is not. But if crews suffered fright, our presence must have put the wind right up the Soviet naval authorities. For unidentified NATO submarines to come close enough to collide with their submarines in their own back yard must have been uncomfortable to digest.

Soviet ships also behaved unpredictably and were quite likely to attempt to ram any submarine they found as *Ugra* had done when I was in *Otter*. *Opportune* was rammed and seriously damaged when inspecting an AGI off the Scillies, the very same operation we had undertaken in *Osiris* off the Mull of Kintyre.

Such operations were so sensitive that some required Prime Ministerial approval, the risk of political embarrassment being considerable. An unfortunate incident at the wrong time could be enough to scupper UK/US/Soviet diplomatic negotiations. They also called for the highest levels of leadership and skill and we had both in abundance. Our Commanding Officers, all having been through the exhaustive Perisher process, were entrusted with virtually full autonomy on such sneaky patrols and, as with SSBNs, would not be signalling for help.

Soviet submarines on the other hand, carried political commissars, used conscripts, and did not spend as much time at sea per capita. Their submarines were neither as well-designed nor as well-built as ours and they were frequently introducing new designs. They even managed to lose about eight of their submarines accidentally, but one should never underestimate the opposition.

Soviet engineers and scientists were exceptionally good. They had developed thermonuclear weapons ahead of us, had put a man into space before us, and way back in 1967, Soviet *Styx* missiles shook the West by sinking the ex-British Israeli destroyer *Eilat* in the first ever sinking of a warship by missile attack. Soviet technology worked. Technologically, they were a formidable foe but the number of submarine accidents pointed to systematic weaknesses.

Kursk, one of their latest cruise-missile firing boats, was only six years old when she had her bows blown off by her own torpedoes. *Komsomolets*, the prototype for a new generation, sank only five years after commissioning. In 1986, the Yankee class *K219* sank after a missile explosion in one of her tubes. And back in 1968, *K129,* a Golf class diesel-electric submarine armed with ballistic missiles, simply disappeared in the Pacific for an unknown reason. As six bodies recovered later had radioactive contamination, missile explosion is one of the postulated causes. All of these submarines carried nuclear weapons. The last Royal Navy submarine loss was the diesel boat, HMS *Affray*, back in 1951. The US Navy had not lost a submarine since *Scorpion* in 1968. Thirty-two years later, the Soviets were still losing their best submarines.

The tragic history of Soviet losses and accidents shows that although their submarine designs had advanced impressively over a thirty-two year period, their safety record had not, and that points to poor safety design, inadequate quality control, weak management, and insufficient training. Of *K129's* crew of eighty-nine, forty were new to the boat on sailing.

Soviet submariners were certainly as brave, aggressive and dedicated as our own, the selfless gallantry of the men in *K19* and *Komsomolets* bearing witness to that. It was the Soviet political system that was letting them down; political commissars ensured that the infallibility of the Party was never questioned. Failure could never be admitted and excuses were found, the sinkings of *K129* and *K219* being blamed on collisions with American SSNs. The Soviet regime was prepared to sacrifice submariners' lives just to show it could match the West.

Similarly, the one-party-state approach to management necessitated cutting corners to meet production targets and militated against the quality of the finished product. The Soviet regime had no democratic accountability. Accidents were simply covered up and the state-controlled media did not report them. Chernobyl was revealed only by radiation monitoring stations in Sweden detecting the fall-out. Who would want to live in a one-party state?

Implicit in our surveillance patrols was the search for any unusual Soviet behaviour such as surge deployments of submarines or the putting to sea of their Northern Fleet. This could either be for a training exercise or a prepositioning of forces for something more sinister, such as preparing for a war.

The second main role for our SSNs was to trail Soviet submarines and study their modus operandi, their tactics, their ability to counter-detect us, and to locate their patrol areas. Such trails could begin up in the Barents Sea and continue down through the Norwegian Sea, through the Greenland-Iceland-Faroes-UK gaps and down into the Atlantic. What was then a top-secret American seabed listening system called SOSUS was another ace in our pack. It was able to detect Soviet submarines at long range and direct our SSNs as well as our maritime patrol aircraft and surface ships, into contact.

The third main role was to de-louse our SSBNs to verify that a Soviet SSN was not trailing them, or to sweep ahead and engage any Soviet SSN that may have been waiting off Ireland for one of our SSBNs to exit the Clyde.

It is frustrating to think that the British public remained completely unaware of all this. The Submarine Service was one of our greatest national success stories. We really were playing at the top of the world's First Division. But not only were the British public unaware; so were most of the Submarine Service. We never discussed our activities except within our own boat and even there few were in the know. Secrecy was essential. That is the nature of the Service. During the two World Wars, for security reasons, the British public were never informed of our staggering submarine losses. In the Second World War, we lost the entire strength of our Flotilla, seventy boats in total, but they had played a major unheralded role in, for example, Montgomery's great victory at El Alamein, because our submarines were preying on the German supply lines from Italy.

****

Whilst in the trail of Soviet boats, SSNs would carry out successful dummy *Tigerfish* attacks using on-board torpedo simulators but these simulated successes, I feared, were creating the false impression that in time of war, we would be able to inflict heavy losses on them. I never subscribed to such optimism. I knew just how unreliable our torpedoes were and felt passionately about it; that was my unfinished business. So I requested appointment as Torpedo Analysis Officer in the Submarine Tactics and Weapons Group (STWG), a new think-tank for submarine warfare development.

At face value this was a backroom job and I would be working for a retired Commander, a chain-smoking veteran of the Second World War. It was a career dead end but it was where I wished to pitch my efforts. We had to improve our weapon system effectiveness. My timing was perfect. The improved *Mod 1* variant of *Tigerfish* had just been accepted into service and it would be for me to analyse its performance. Marconi had won the contract for design, manufacture, and proving trials of the weapon. They had even been contracted to write its Acceptance Trials Report and that aroused my suspicion. I read the Report in detail. I was one of the few in the Navy to do so and my worst fears were confirmed. The Executive Summary, the only bit that most people read, was a masterpiece in half-truths. It claimed that in thirty-two test firings, the weapon had demonstrated all its Agreed Characteristics. That was true but it did not mention that it had never achieved a completely successful run. It was pick-and-mix reporting, like describing a pearl necklace bead by bead but omitting the fact that the string kept breaking. My task was to establish the truth. I had to analyse every practice firing by our submarines and assess their true effectiveness in real action.

In this endeavour, I inherited the all-female team of six highly intelligent young Wren Analysts first established in the original Torpedo Trials Unit. The team was now led by Jan Martin, a vivacious young Second Officer with a First Class Honours degree from Edinburgh University, who revelled in reminding me that her university was superior to mine. I doubt if a single sailor in any of my submarines had A-level maths but all of my girls did and that brought home to me the vast untapped pool of female talent that would be available if the Navy were to open its doors to women.

I did not have to look far for evidence of the new weapon's unreliability. There was the usual crop of guidance wire breakages at launch, just as I had seen eight years before, and when the weapon was successfully guided to its target and homed in for the kill, it suffered consistently from near misses. From memory, I reckoned the chances of a successful attack at no more than one-in-five. It was fundamentally a superb weapon but not yet fit for purpose. Its faults had first to be admitted and then eradicated.

I don't know if some cosy, high-level, blind-eye-turning deal had been struck between Marconi and the Ministry of Defence Procurement Executive, but either way, accepting *Tigerfish Mod 1* into service was, in my eyes, a deception. As long as we did not go to war, it was not a problem. With stalemate in the Cold War, real underwater dogfights seemed highly unlikely but it was no way to run a Navy.

The *Mod 1* had been designed to attack surface targets as well as submarines, but not to hit them. It was designed to pass close beneath, have its magnetic fuse triggered by the target's hull and thus break the victim's back. In this it proved spectacularly successful. Indeed, when we fired one against the tethered hulk of an old frigate, it broke the ship in two. Against a moving target though, it kept missing astern. The reason, I diagnosed, was because it had been designed to level out at a pre-set shallow depth and as it dipped its head on to this ceiling when homing up from deep, the target disappeared from its listening beams and it stopped homing. It was as if it had caught a rabbit in its headlights and swerved at the last minute to avoid the target. It would barely miss but a miss is as good as a mile. There was a clear design fault in the homing logic.

Marconi disputed my analysis but the facts were irrefutable and I won my case. It was then agreed that a design change should be introduced in a *Mod 2* variant – a further contract for them. However, even without modification, the *Mod 1* could still successfully attack a surface target if fired 'up the chuff' and not from the classic broadside position. Then, the weapon would continue to run on under the target even if it did go deaf at the last minute. This called for a change in tactics and I was in exactly the right place to influence that change. As our SSNs had perfected the art of tucking in behind Soviet submarines and trailing them, attacking from the target's blind stern arcs created a perfect synthesis between approach and attack.

However, not all failures were due to *Tigerfish*. The somewhat self-assured Captain of one SSN whose firing had been unsuccessful was fulsome in his condemnation of the weapon but my analysis revealed that it had been guided into a mud bank. With its head buried in the mud, the torpedo had continued to obey all his commands but went nowhere. It was simply wagging its tail for him.

'You've got your analysis wrong, Eric,' he declared, convinced of his ability to walk on water.

'No I haven't.'

'Yes you have.'

I was then summoned to justify my analysis to the Squadron Captain. I showed him the track of the weapon superimposed on a chart of the Sound of Raasay. It had been guided into a shallow patch. Alas, our torpedo guidance displays were too primitive to display seabed topography.

****

Even in peacetime, submarines carried a war load of torpedoes so that they could be deployed immediately should war suddenly break out. *Conqueror* happened to be the first to carry a war outfit of *Tigerfish Mod 1* torpedoes and conducted the first two warshot proving firings. The warshot had a number of key differences from a practice weapons, the most important being that it had a warhead and not a data recorder. The warshot also had more than twice the range of the practice variant, about forty kilometres, and had a much more powerful battery that was primed only at launch. Both these warshots were duds: one being a stillbirth, its battery failing to prime, and the other suffering the familiar breakage of its guidance wire at launch. With this one hundred per cent failure rate, one could hardly blame *Conqueror* for having no confidence in *Tigerfish*.

When contemplating war, the maxim is always: *expect the unexpected*. In March 1982, the unexpected happened. Argentina invaded the Falkland Islands and the Royal Navy went to war not with the Soviets but with a modern Western navy. The Argentines were equipped with German-built submarines, British-built ships and with American, French and British weapon systems. Our teeth would be put to the test.

By coincidence, *Conqueror* was then at Faslane and Commander Chris Wreford-Brown had just relieved my old term-mate from Dartmouth, Commander Richard Wraith, as Commanding Officer. (Richard's father, also a submarine Captain, was lost at sea when his boat was sunk in the Mediterranean at the same time as my father's ship). *Conqueror* was ordered to store for war on 31st March and sailed for the South Atlantic on 4th April.

Once in area, she found the elusive Argentine cruiser *General Belgrano*, equipped with lethal French-made *Exocet* anti-ship missiles, and her two escorting destroyers. The group presented a serious threat to the British Task Force under the command of Admiral Sir 'Sandy' Woodward, a fellow submariner. The Task Force included two aircraft carriers, *Hermes* and *Invincible*, and thousands of assault troops in various passenger liners. The loss of any of these major units would have had the most serious consequences for the re-taking of the Falklands and so Admiral Woodward called immediately for the *Belgrano* to be sunk. After consultation at Cabinet level in London, the signal '*Sink the Belgrano*' was sent.

The eyes of the Prime Minister, Chief of the Defence Staff, First Sea Lord, Commander-in-Chief of the Fleet, Flag Officer Submarines, and Sandy Woodward were now all on Chris Wreford-Brown in *Conqueror*. Chris knew the *Belgrano* was not going to hang about as a sitting duck. This was what he had trained for thirteen years to do and was exactly the situation for which *Tigerfish Mod 1* had been designed. Chris, however, had no faith in *Tigerfish* and was not prepared to risk the mission because of dud weapons.

Thus the *Belgrano*, an ex-American Second World War cruiser, was sunk in a classic periscope attack from 1,400 yards range (1.28 Km) using a salvo of Second World War vintage, straight-running, diesel-driven torpedoes. It was like a historical re-enactment but for real. The lack of integrity in accepting *Tigerfish* into service had been thoroughly exposed.

In a copybook attack, Chris Wreford-Brown scored two direct hits and the *Belgrano* was sunk. *Conqueror* then made a fast getaway and as she did so, depth charges dropped by *Belgrano*'s escorts could be heard exploding. She had been in harm's way.

It was the first time in history that a nuclear submarine had sunk an enemy ship. *Conqueror* had travelled submerged from Faslane to the opposite end of the world before sinking a dangerous enemy cruiser and, in so doing, confined the entire Argentine navy to its own territorial waters for the rest of the war. Their aircraft carrier, *Veinticinco de Mayo*, the greatest enemy threat, never ventured out of Argentine territorial waters. Had she done so, she would have been sunk with even greater loss of life. By then we had three nuclear-powered submarines, *Splendid*, *Spartan* and *Conqueror*, in area with *Valiant* on the way.

The sinking of the *Belgrano* cost 321 Argentine lives and some at home protested about Britain's action with facile arguments such as: the *Belgrano* was heading away from the Task Force when she was sunk. She could, of course, have altered course at any time. Thirty years later, *Belgrano*'s Captain admitted that his mission had indeed been to attack the Task Force. Of course it was! This was a shooting war. I analysed *Conqueror*'s attack and awarded it full marks.

Chris Wreford-Brown had fearlessly taken his boat into very close range to be sure of success. A much longer range *Tigerfish* attack would have significantly reduced the risk. Hypothetically, a *Tigerfish* could have been launched from astern of the *Belgrano* at, say, six thousand yards (or more), which was a much safer firing position with far less need to expose a periscope and risk counter-detection. As *Belgrano* was travelling at twelve knots, a *Tigerfish* in search mode would have caught her in fifteen minutes and with *Belgrano* being a noisy target, the torpedo would have homed automatically from over a thousand yards. It would have homed on her propellers, blown off the stern and completely disabled her with possible sinking as a result. In such an attack, *Belgrano* would not have had the faintest idea of *Conqueror*'s whereabouts. That is how it should have been, but a hot war attack is not the place for experimenting with unreliable weapons. Chris Wreford-Brown's choice of the tried-and-trusted *Mk 8* torpedoes was absolutely correct. Working weapons win wars, not weasel words.

On hearing of the sinking, a Marconi rep was on the phone immediately for confirmation that *Tigerfish* had been used. That would have worked wonders for its export potential. (We should have lied).

The surface Navy loves ceremonial; the Submarine Service does not. The latter's ceremonial activities are confined almost entirely to a handful of Squadron staff meeting their boats on return from patrol, more akin to meeting a child bringing a medal home from school than Trooping the Colour. When *Conqueror* returned three months later from her thirty-thousand-mile war patrol, the only concession to ceremonial was the flying of the Jolly Roger, the pirate flag, and that was purely for the benefit of the small group of Squadron officers waiting on the jetty. The incident hit the headlines though, and the Navy took some flak for this apparent display of heartless triumphalism.

Flying the Jolly Roger on return from war patrol is not actually bravado but a long-established submarine tradition begun in the First World War by Lieutenant-Commander (later Admiral Sir) Max Horton when in command of submarine *E9*. On return from a patrol in which he sank the German cruiser *Hela*, he flew the Jolly Roger but not as a display of triumphalism. It was his subtle way of waving two fingers at Admiral Sir Arthur Wilson, the First Sea Lord, who had stated that submarines were: *'underhand, unfair, and damned un-English,'* and that he would: *'have the crews of captured submarines hanged as pirates.'*

Conqueror's Jolly Roger was unique. It was the first nuclear submarine ever to sink an enemy ship and thus its flag bore both the symbols of an atom and a warship, the crossed bones being replaced by crossed torpedoes. It also bore the symbol of a dagger meaning that she had been involved in other covert operations such as landing Special Forces in South Georgia and sitting off the Argentine coast providing early warning of aircraft taking off to attack the Task Force.

****

Margaret Thatcher, Prime Minister during the Falklands War, duly emerged as a superb war leader, very much in the Churchill tradition. She had put her faith entirely in the country's military leaders, in particular its Naval leaders, for this was a long-range, amphibious war. As it happened, the then Chief of the Defence Staff was Admiral Sir Terence Lewin, who had been Captain of the Dartmouth Training Squadron when I was a Cadet. Admiral Sir Henry Leach, First Sea Lord, had been our VIP visitor in *Revenge* on return from my steam leak patrol and I had taken him on a personal tour of the Engine Room. I had discussed torpedoes with Admiral Sir John Fieldhouse, Commander-in-Chief of the Fleet, when he visited STWG. Sandy Woodward, in Command of the Task Force, had been Captain of *Warspite* when I was in *Otter* and I would later work directly for him. I had met our entire top chain-of-command from Prime Minister down to submarine commander. The submarine COs were of my own vintage and I counted them all as friends: Roger Lane-Nott (*Splendid*), Tom le Marchand (*Valiant*), Chris Wreford-Brown (*Conqueror*), James Taylor (*Spartan*), Rupert Best (*Courageous*) and Jonathan Cooke (*Warspite*), the latter two having joined the Navy with me. It felt strange and somewhat frustrating to know all the main players in a real war but to be no more than a spectator.

Whilst Commanding Officers rightly earn the glory, one must not forget the achievements of their engineers. For such technically complex machines as nuclear submarines to operate independently on war patrols of up to three months duration at the opposite end of the world was also a *tour de force* by them.

Onyx, under the command of Lieutenant Commander Andy Johnson, was the only diesel submarine to deploy to the Falklands and had the longest patrol of all on account of her very long passage time both ways. It was almost four months in total which I think deserves a medal. She was specially adapted for inserting Special Forces and she also carried *Tigerfish* torpedoes. In both roles, she had bad luck. In going close inshore for a covert landing of a Special Boat Service team, she struck an uncharted rock and damaged two torpedo tubes, trapping one *Tigerfish* in its tube. Unknown to *Onyx*, its battery had primed, meaning it could have exploded at any time, but that was only discovered on return to Portsmouth.

Onyx was also charged with sinking the hulk of the ill-fated RFA *Sir Galahad*, which had been wrecked by Argentine aircraft with great loss of British life. She fired two *Tigerfish* at the hulk and both failed. For the second time, our old *Mk 8* torpedoes had to be used to do the job.

On the back of Britain's spectacular victory, Margaret Thatcher was swept back to power in the following General Election. The lady had certainly done well but it was her Government that had, in effect, caused the war by withdrawing the Falklands guard ship as a savings measure. That had led the Argentine dictator, General Galtieri, to think that taking the virtually undefended islands would be easy. He was right. But holding on to them was a different proposition. As the Falklands lay four hundred miles across the South Atlantic from Argentina, Galtieri needed a navy, first to land his invasion force, and then to supply it and prevent counter invasion. He had landed almost nine thousand troops by the time the British Task Force arrived. In sinking *Belgrano, Conqueror* had neutralised his navy and isolated his army.

At the start of that war, Britain had advised all submarine operating nations to keep their submarines out of the combat zone, as they would be attacked as hostile. As a far as I know, the Soviets complied but they did take advantage of our migration from the North Atlantic. They deployed seven of their own submarines there during the conflict, but we had not dropped our guard against them. Our remaining SSNs were continuing, along with the US Navy, to maintain surveillance patrols against them, and our SSBNs were continuing with their Strategic Nuclear Deterrent patrols. The Submarine Service was at the very top of its game.

The Falklands War demonstrated that whilst Strategic Nuclear Deterrence prevents world war between global superpowers, it does not prevent minor wars involving lesser forces. *Conqueror*, however, had demonstrated a sub-nuclear form of deterrence. She had shown the Argentine navy that its destruction was assured if it dared to attack us. Had a nuclear submarine been deployed to the area before Galtieri's invasion, the Falklands war could have been avoided. The war was caused by failures in Intelligence, political misjudgement, and the lack of a pre-emptive show of force.

****

In the immediate aftermath of the Falklands War, it was decided that a captured Argentine ship should be used for target practice, first by aircraft from our carriers, then by gunfire from various ships, and finally in a *Tigerfish* attack by one of our SSNs. By the time the submarine was given clearance to fire, the target was already sinking and as a dead ship makes no noise, the attack had to be carried out with the weapon in its active mode, i.e. pinging and looking for return echoes. In this it was highly successful. It quickly found its target and conducted a perfect homing attack, but failed to detonate and lost contact. It was then guided round manually for further attacks, all with the same result – perfect homing but no detonation.

The firing came back to me for analysis as yet another failed *Tigerfish* attack but one should never jump to conclusions. This weapon had not failed. From the sequence of attacks it was clear that the 'dead-ship' target had been on the move and travelling at speed. Were there ghosts on board? The explanation was that ships do not sink vertically like stones but ski downhill to the deep ocean floor. This target would travel two miles before it reached the seabed. What the torpedo had been attacking was the massive trail of bubbles streaming out behind it. Bubbles in water reflect sound like solid objects – ask a Humpback whale – but bubbles do not have a magnetic signature and could not trigger the torpedo's fuse. Against a live target, that *Tigerfish* would have achieved a kill.

In conducting this analysis, I shuddered at the thought of a warshot torpedo being turned back for a second attack; torpedoes are not designed for second attacks. If the guidance wire were to have broken while it was heading back towards its parent submarine, there would have been no way of stopping it. It could have been a submarine version of an own goal.

****

The crowning glory of my time in STWG was when I took my harem of Wren analysts to AUTEC, the Atlantic Underwater Test and Evaluation Centre, a deep-water American torpedo tracking range in the Bahamas. There, for the first time, *Tigerfish* would be tested in its intended role as an anti-submarine torpedo by actually attacking a deep-dived submarine, something we could not do in the relatively shallow waters of our own tracking range in Scotland. Two of our SSNs were to conduct tracked dogfights with each other with both tactics and torpedo performance being assessed.

For a torpedo analyst, AUTEC was paradise. The analysis facility was on the beach. American helicopters would recover our weapons immediately after each run and we would be crunching the data within a couple of hours of each firing. Better still, we could watch each attack unfold on a large tracking screen and for icing on the cake, between firings, I could retire to the beach with my six bikini-clad young Wrens.

The crowning moment was when our two submarines approached from opposite ends of the range, each out of sonar contact with the other. One was playing the role of a British boat with a towed array (now a standard fit since the days of the trial in *Osiris*) while the other was simulating a Soviet submarine using only hull-mounted sonar. The towed array gave a much longer detection range but required a lot of manoeuvring to refine the target's position whereas the hull-mounted sonar gave shorter-range detections but more precise target information that allowed a speedier strike. The burning question was: which would have the upper hand?

Watching the drama unfold was as good as any thriller movie. I watched in suspense as the dotted tracks of the two submarines appeared at opposite ends of the screen like two gladiators entering the arena, but I could only guess at what was happening on board. I watched the towed array boat make long zigzags; that meant she was in contact and trying to set up an attack. The other boat simply tracked towards the centre of the screen in a slow curve which suggested he was still searching. Then, almost simultaneously, secondary tracks appeared from both submarines. They had both launched torpedoes. That was not in the script. What would happen when the torpedoes met? Would they attack each other in a torpedo dogfight? No, they simply passed in silence and continued to their respective targets. Both torpedo tracks then merged with their respective targets. I could scarcely wait for the data recorders to be recovered.

Both torpedoes had homed successfully and both had gained fuse actuations. It was a one-all draw. The submarines had just killed each other. For *Tigerfish*, it was a sensational result and a complete vindication of the investment in that weapon. In its true anti-submarine role, *Tigerfish* was a winner. The Americans were hugely impressed. After that we moved up a gear and conducted similar exercises with each submarine controlling two torpedoes simultaneously, which, by then, our control systems could do.

When the residual reliability problems, mainly to do with launch, had been resolved, *Tigerfish Mod 2* became the best torpedo in the world because it was a weapon of stealth. At last, our SSNs had guaranteed teeth. It had taken only twenty years to get *Tigerfish* right, thirty from initial concept, but now we really could play the silent assassin. Good timing as the Soviets had now developed torpedo decoys. Stealth was of the essence.

The Soviets had also changed their maritime strategy. Now, instead of trying to emulate NATO navies, they were going for a Fortress Russia philosophy. They were concentrating their submarines closer to home, defending in depth, and relocating their SSBNs under the Arctic ice cap for added protection. Their gargantuan new Typhoon Class, the one featured in the film *Hunt for Red October*, all twenty thousand tons of it, was big enough to smash through the ice cap to launch its missiles. That was a true game changer; submarine warfare under the ice cap is a different proposition. Apart from the creaking and groaning, ice keels hang down like massive stalactites and would create confusing reflections from any torpedo sonar transmission. *Tigerfish*, however, had good passive homing capability and that gave it potential for under ice operations. The Americans even considered buying some.

****

AUTEC was managed for the US Navy by the RCA Company and was restricted by the Bahamian Government to a fenced-in compound in which no building above two storeys was permitted. It was more like a work camp than a tropical paradise, with even the most senior RCA managers living in caravans. As several thousand male Americans worked there with only a handful of females and virtually no married-accompanied personnel, my clutch of attractive young Wrens in their white tropical uniforms became an instant attraction. I felt more like a father with six vulnerable daughters than a lucky old man with a private harem.

One day, a senior RCA man by the name of Chuck, invited me to join him in a snorkelling expedition on the coral reef and to bring Sue Wheeler, Jan Martin's successor, and one other Wren with me. Clearly he was either being politically correct with his gender balance or saw me as a procurement agent.

As the reef lay only half-a-mile off the beach, I had no concerns when we set off at six o'clock in the morning in only tee-shirts and swimming costumes, but we were not going to the visible reef. We were to spend an hour battering along in a high-speed open boat until well out of sight of habitation. As Chuck had invited the ladies to sit with him in the stern, it fell to me sit in the bow to balance the boat. I found myself being showered with cold spray every few seconds as the boat pounded into the short waves. At first I tried to act tough but not for long. After a quarter-of-an-hour I was damned cold. After half-an-hour, I was damn near hypothermic. And by the time we reached Chuck's selected spot on the reef, I was a teeth-chattering, shivering wreck and had lost all interest in snorkelling. Chuck and the girls, all dry as bones and basking in the morning sun, were rarin' to go.

When we reached our destination, Chuck shouted: 'Throw the anchor overboard, Eric.' Before I could move, he was over the side and paddling off across the reef. It was a prime display of American virility. Seconds later, the girls were over the side and following him.

We were about a mile off an uninhabited, palm-forested shore yet, only fifty metres away, beyond the reef, lay thousands of miles of open Atlantic. Apart from feeling hypothermic, the very idea of heaving the anchor over the side and abandoning ship ran counter to all my instincts. Rock is not a reliable anchorage and the wind could cause the boat to drag its anchor and it could drift off into the open ocean. 'I'll stay here as safety number,' I shouted like a boring old fart, but they didn't hear me. They were face down and paddling away with alarming speed.

After a few minutes of shivering in the boat, I decided to pull myself together, donned my flippers and snorkel, jumped over the side and almost drowned. I was a good swimmer and a qualified SCUBA diver but had never before jumped off a boat wearing a snorkel. With an aqualung, one continues to breathe normally through the mouthpiece and I did that instinctively. With the snorkel, I sucked in a lungful of seawater. I was in immediate trouble, gasping for air and in danger of drowning. In blind panic, I reached up to grab the boat but it was bobbing about and I was under a high side. I could not reach the top lip and there was no one to pull me up. One finds amazing strength when desperate. Somehow I managed to kick up, grab the gunwale and haul myself back on board where I flopped into the scuppers, gasping like a freshly caught fish.

Feeling utterly stupid, I gathered myself for a second attempt and made exactly the same mistake. When the others returned, they had no idea that I had been dicing with death.

'Let's go over there,' Chuck shouted, pointing at a dark patch in the lagoon; and off we went.

This time, I was determined to rescue my pride. As we were inside the lagoon, I had less concern about safety and decided to show my mettle by being first overboard. I was already in snorkel and flippers and perching on the side of the boat before Chuck had even cut the engine but I was not going to make the mistake of jumping and instead slipped gently into the water. But this time the right leg of my swimming costume caught on a nail and as I descended into the sea, my costume was ripped off. For the second time in twenty minutes I was desperately reaching for the lip of the boat but this time in the nude.

As Chuck was already over the other side and paddling off on another voyage of discovery, I was left hanging on to my side, trying to protect my modesty, with Sue and Margaret, one of my Analysts, grinning down at me. They could not believe their luck. They really had caught the boss with his trousers down. Sue loved to tease me at the best of times and was now having a field day. 'Where's your costume, Ericky?' she called, her face a picture of delight.

I was far too shy to haul myself back on board wearing only facemask and flippers. 'Pass me down my jogging shorts,' I called. 'They're in my bag.'

'Oh, I don't think we could go into Sir's bag, Margaret?' Sue asked rhetorically. 'Never know what we might find in there.'

'Just pass me my shorts please.'

'Oooh, Ericky. We just need to take some photographs. Don't we, Margaret?'

'Never mind bloody photographs, just pass me my shorts. My arm's coming off.' I was hanging on to the boat with one arm and it was bobbing up and down.

When eventually my shorts were passed down, I found that I could not put them on whilst wearing large flippers, so the flippers had to come off and that made it even more difficult to keep hanging on. By now the girls had their cameras out.

Wrestling to remove flippers and then pull on shorts with only one hand whilst bobbing up and down by several feet and having my other arm stretched to near breaking point proved damn nearly impossible. Eventually I succeeded but only to find that the shorts would not come up over my crotch; I had pulled them on with the legs twisted in a figure of eight. That was the last straw. I let go of the boat, swam free, pulled down the shorts, untwisted them, and pulled them on again with the girls clicking away. I would not be allowed to forget this. In future, they could do without me as a chaperone.

****

As my summer leave was taken up with Scout camps, we took our family holidays in winter. In January 1983, while skiing in Austria, our hotel receptionist handed me a note. It was a message from my father: could I phone home please? My heart hit the floor. Something must be seriously wrong. Father never called me on holiday. Had Mother died? I phoned home immediately.

'Is there another Lieutenant Commander F G Thompson in the Navy?' he asked.

'No.'

'Well, someone by that name has been listed in the Daily Telegraph January Promotions.'

I was now a Commander.

CHAPTER 19

The Joint Service Defence College

'They are an abomination; they shall not be eaten.'

BIBLICAL QUOTE ON THE CORMORANT

When the Commandant welcomed the inaugural course to the new Joint Service Defence College (JSDC) at Greenwich by proclaiming that we were 'handpicked men of the future, destined to roam the corridors of power,' I smiled. I had requested a foreign, married-accompanied, fun-in-the-sun appointment but none was available. 'Then how about a staff course?' I said to my guilty-looking Appointer. He took a hasty look into his crystal ball and mumbled grudgingly: 'If you can start next week, there's a vacancy on JSDC1.' That was why I was there.

I was a newly promoted Commander, a complete stranger to the world of military pomp, and my staff skills were unrefined. I had spent my entire working career in the closed world of submarines. My new comrades-in-arms came from all walks of the British military establishment: pilots, pay-wallahs, logisticians, cavalry, infantry, police (civil and military), paratroopers, ship and submarine captains, Intelligence Corps, Royal Marines, Royal Artillery, Civil Service (Foreign Office and Defence), and the Deputy Chief Executive of Gatwick airport. We even had a real spy. I struck an immediate rapport with Wing Commander Peter Johnson RAF, a Jaguar pilot, and Chief Superintendent Ernest Bleakney from the Royal Ulster Constabulary, who were both fellow front line operators.

The College crest was a Cormorant, a creature that could dive like a submariner, paddle like a skimmer, waddle like a soldier, and flap like an airman, so we students were known as Cormorants. We were to be indoctrinated in the virtues of joint operations to break the tradition of inter-service rivalry and regimental introspection.

It was September 1983 and the Falklands War had been the previous year. Chris Keeble, our paratrooper, had been Second-in-Command of 2 Para at Goose Green where Colonel H. Jones had won his posthumous VC for leading the charge against a nest of Argentine machine guns. I remembered having seen Chris, distinctive by his prematurely white hair, on the TV news of the time, leading the burial service for his fallen Commanding Officer and fellow soldiers on the field of battle. Neil Thomas had been an ace *Sea Harrier* pilot in the conflict and was a survivor of dogfights with the Argentine air force. His war cry was: *never give a sucker an even break.*

The Troubles in Northern Ireland were also still in full swing and Ernest Bleakney was fresh from the embattled Springfield Road police station in West Belfast, one of the hottest spots in the City and frequently attacked by the IRA. Many others had also seen active service in one or both of these conflicts and most of my new Army and RAF colleagues had served with the British Army of the Rhine, facing the Red Army. Compared with such battle-hardened colleagues, I felt a bit of a fraud. I had been tucked up safely in my submarines at risk only from material failure or our own cock-ups. However, I had been awarded an MBE and was one of only eight (out of fifty-two) students who had been decorated.

The Joint Service Defence College stood opposite my old alma mater, the Department of Nuclear Science and Technology, and, as if to make up for the modesty of the latter, was not short on pomp. At the inauguration ceremony, the guests ranged from Field Marshal Sir Edwin Bramall, Chief of the Defence Staff, to the Mayor of 'Nuclear Free' Greenwich, resplendent in his gold chain and CND badge, seemingly unaware that a nuclear reactor lurked in Sir Christopher Wren's basement. It was all a far cry from sweaty engine rooms.

'May I introduce my incidental attraction, Sir?' stammered one of the army officers at the College opening ceremony. In introducing his wife to the Commandant, he was trying to obliterate any impression that another woman in his life might rival Her Majesty.

Later, a tutor confided that bachelor officers in the Army had no chance of commanding a regiment. In the Army, one's wife was part of the promotion calculation. She was required to act as Queen Bee amongst the regimental wives and therefore had to be socially acceptable. I wonder how that works in today's era of gay marriage and equal opportunities?

The Course lasted six months and had a superb syllabus. It included: lectures from very senior military officers, top civil servants, leading politicians of both main parties (including the Minister of Defence Michael Heseltine MP), prominent members of the public and business sectors, and top academics. We studied everything from NATO defence strategy through terrorism to national energy requirements. We learned about the carpets of tactical nuclear weapons that would be laid down in East Germany if the Soviet horde ever decided to roll into West Germany. We visited the British Army of the Rhine, the depressing remains of the Bergen Belsen concentration camp, the reality of the Berlin Wall and, when we passed through it via Checkpoint Charlie, discovered the bleak streetscape of East Berlin devoid of retail outlets and traffic.

Simply by passing through the Wall, we had moved from the affluent boulevards of West Berlin into the grim austerity of the Communist world. There was no obvious sign of extreme poverty or starvation but we had entered a different world, a police state, a barren land devoid of freedom in which even members of one's own family could not be trusted. People were shot simply for trying to move from one side of the city to the other. We were even taken to the supposedly prestigious GUM department store, the high temple of Soviet retail therapy. Compared with Harrods, it seemed more like a government surplus store in downtown Glasgow. We had gone back seventy years into the world of the Bolsheviks, to a society living in a war economy.

We also visited the Berlin Air Traffic Control Centre, still operated by the four victorious military powers as if the Second World War had just ended. Soviet, American, British and French military personnel still working together as Allies but actually man-to-man marking each other.

As a submariner, I had only imagined the enemy. Now I was seeing it in its natural habitat and I felt not a shred of animosity, only a deep sense of pity. I felt more inclined to help than fight. But the misery of East Berlin was not our fault; it was the Communist regime that was the cause of their misery (and before that, the Nazis). It is the political regime that is the enemy, not the man in the street or the soldier in the field – and nothing is more intent on saving its own skin than a political regime. That is why the Strategic Nuclear Deterrent works.

At the end-of-course dinner in the fabulous Painted Hall, one of my fellow students had organised a spoof prize giving and, to my delight, I was awarded the Communications Prize: 'for boring the pants off everyone with my diatribes on failures in procurement of the *Tigerfish* torpedo.' My prize was a goldfish in a polythene bag procured from a local funfair. I was thrilled. This was something to cherish. And to ensure the polythene bag did not burst, I borrowed an empty ice bucket from the bar, took 'Tigerfish' up to my cabin and left him on my desk swimming happily in the ice bucket.

When I returned hours later, poor little 'Tigerfish' lay dead on my desk. Like his namesake, he had suffered a problem with shallow ceiling control and leapt out of the bucket, a grim metaphor.

As I had three fallow months before taking up my next appointment, I requested secondment to the Admiralty Underwater Weapons Establishment (AUWE) at Portland to find out what was in the research pipeline for future torpedoes and to extend my university thesis on the auditory systems of bats, cats and humans into those of marine mammals. The scientists there were wonderfully helpful and gave me full access to all their research, which allowed me to develop a revolutionary design concept for a new torpedo, modestly christened *The Thompson Torpedo*.

The lay reader may not immediately associate Killer Whales with submarine warfare but there is direct comparability as both rely on sound. Killer Whales, submarines and torpedoes are all in the same business, but Nature's supreme predator has been perfecting its technique for millions of years while nuclear submarines and homing torpedoes are less than eighty years old. There must be, I reasoned, a lot we could learn from Killer Whales.

The hearing of a whale is a mystery, but there is a clue in that we are both mammals. For example, we can tell the difference between a symphony orchestra and a rock band, or tell when something is wrong with the car because of a new noise. Yet, we have only two simple eardrums whereas submarines and torpedoes have complex arrays of listening devices. The secret of our phenomenal hearing lies in the processing power of the human brain, the most sophisticated sound analyser in existence – with the possible exception of Killer Whales and dolphins.

The question is: how do Killer Whales hear underwater? They have no obvious ears. Millions of years ago, their ancestors did because they were land animals before going back into the sea. Killer Whales actually have airborne hearing systems like our own but have lost their external ear flaps because their bodies have been streamlined for high underwater speed. Had external ears remained, they would have heard nothing when hunting at high speed; it would be like driving fast in a car with the windows open. They have, however, retained eardrums like our own but these are buried deep inside their body. As whale body tissue has the same acoustic density as water, sound simply passes through it to reach their eardrums.

What is even more sophisticated is that they can still tell exactly the direction from which a sound is coming by measuring the microscopically small time delay between left and right ears.

My conclusion from this was that the *Tigerfish* torpedo, though advanced for its day, was primitive in the extreme compared to a Killer Whale. The *Thompson Torpedo* was an attempt to reach parity. It would have a streamlined, teardrop hull and not the parallel-sided cylinder required to fit into a twenty-one inch torpedo tube designed before the First World War. My torpedo would have listening devices buried in simulated flesh, not screwed on to a rigid stainless steel plate on its nose. It would have a frequency range extending from the low rumble of machinery to the highest pitches of ultrasound, each frequency band playing a different role in the hunt. With *Tigerfish* technology, we were in the somewhat ludicrous situation of the submarine detecting its target at low frequency and then sending out a torpedo to find it in high frequency. It was akin to the submarine tuning in to Radio Four Long Wave and expecting the torpedo to find the same programme on Radio One FM.

Nor would my torpedo be fired like a bullet from the bow; it would be slipped out silently from the rear like laying an egg. It all seemed so obvious – but it would require a major re-design of our submarines. That raised the classic chicken-or-egg argument: which came first? Does one design the torpedo to fit the submarine or design the submarine around its weapon?

On completion of my study, I was appointed to FOSM's staff as Signature Reduction Officer, the lead officer in stealth technology. My new boss was Admiral Sir Sandy Woodward, the Falklands victor. He was a highly intellectual Admiral and, at risk of sounding presumptuous, seemed to be on a similar wavelength. He once remarked to me: 'If you don't have self doubts, you're a fool.' I thoroughly agreed.

When he heard of my *Thompson Torpedo*, he asked for a presentation and was suitably impressed. When I said that I thought it was all pretty obvious, he replied: 'What's obvious to you, is not obvious to others.' I took that as a compliment.

I was then summoned to Whitehall to repeat my presentation to the Naval Staff. Though I say it myself, I had a class act. It was well polished, clearly illustrated, and delivered with conviction in best Whitehall-speak. I had absolute confidence in my theory and was well able to defend it against any challenges from the floor since I had already run it successfully past the scientists.

When I returned to Northwood, I received a telephone call from a recently retired Commander, now working for Marconi: 'We would very much like you to come and have lunch with us and tell us all about your new torpedo concept,' he said in his most come-hither voice. All very flattering but I was not going to trot along to a commercial company and give away all my secrets for the price of a free lunch. 'If you want to hear my presentation,' I replied, 'please make a formal approach through correct Ministry channels.'

The following morning, my phone rang. It was the Torpedo Desk Officer in the Ministry of Defence. 'Have you just accused Marconi of making an improper approach?'

'Yes.'

'Thompson, you're a cunt.' (his exact words) 'The Chairman of Marconi has just been on to the Controller of the Navy to protest and my boss is not best pleased.'

His boss, the Director of Naval Operational Requirements, was so displeased that later he would blackball me for an appointment to his department.

The triumph of Ministry politics over technical merit was now apparent. The Defence Procurement Executive in tandem with Marconi was busy making the case for *Spearfish*, the successor to *Tigerfish*. This would be a major weapon project following the traditional principle of making the weapon fit the submarine and it would have significant impact on the Defence budget. The last thing the Ministry needed right then was some bright spark called Thompson promoting an alternative concept. This was the machinery of Government in action. Once a course has been set, it is nigh on impossible to change direction. There are too many reputations at stake, and too many careers and business interests need to be protected. But that's reality. Thus, my brilliant concept was kicked into the long grass, in effect forever; a new torpedo project would not come along again in my career lifetime.

I was not surprised. My torpedo would have required a new class of submarine.

CHAPTER 20

Stealth

Silence is golden.

ANCIENT PROVERB

A hound does not hunt by sight. It sniffs. It looks for the unique aromatic signature of the fox and disregards all other scents. In underwater warfare, the principle is the same. As an enemy submarine cannot be seen, its unique sound signature becomes the target. Conversely, to prevent counter-detection, reduction of one's own sound signature to vanishing point is the ideal. In my new role as Signature Reduction Officer, I was to take the lead in developing the Submarine Flotilla's stealth techniques. My job was to be the link between cutting-edge scientists, submarine designers and front-line submariners. I had also to be aware of what Intelligence was discovering about Soviet submarine signatures. I was at the very core of the underwater warfare business.

A submarine's signature is highly complex. It can include noise, magnetism, infra-red radiation, electro-magnetic radiation, radioactivity, radio and radar transmissions, wake, gaseous exhaust, not to mention sewage and phosphorescence. As the target for active sonars, the hull acts as an underwater reflector, as do the periscopic masts for radar. I had to look at reducing or eliminating all such signatures. Ranged against us were a variety of detection systems: other submarines, ships with active sonars, passive towed arrays and electronic detection devices, aircraft and helicopters with magnetic anomaly detectors, gas samplers, air dropped sonar buoys and homing torpedoes, seabed listening arrays, shore radio intercept stations, and satellites.

With us were the Admiralty Research Establishments at Portland, Teddington, Slough and Portsmouth, where our Government boffins were maintaining the great British tradition of brilliance and eccentricity. We also had the nuclear submarine design sections in Bath led by the illustrious Royal Corps of Naval Architects, aided and abetted by some of the best commercial brains in the country. These included highly specialist companies like Rolls Royce and Associates and the Yarrow Admiralty

Research Department in Glasgow. This was UK Limited's first eleven. We were not at the top of the world submarine league by accident.

It was all highly classified work, much of it Top Secret UK Eyes Only, as information on our submarine signatures was not exchanged, not even with other NATO navies, and would be very highly prized by Soviet Naval Intelligence. Everything in my career so far had prepared me for this and my Masters degree in Acoustics gave me added credibility with the scientists and designers. But in my own submarine community, I was a round peg in a square hole. Signature Reduction spanned all submarine departments but belonged to none. I was in many respects a lone wolf falling into neither the Engineering nor the Seaman branches, a management issue I would have to address.

Since the end of the Second World War, the Royal Navy had been a front-runner in the development of quiet submarines and, since the inception of nuclear submarines, we had enjoyed a significant detection advantage over the Soviets. Now we had advanced to the stage of diminishing returns, each small improvement requiring more and more effort. On the other hand, the Soviets were able to make substantial improvements with much less effort from their lower starting point. The detection gap was closing. In some areas, they were even in the lead.

To put my new role into context, this was the year that American author, Tom Clancy, published his best-selling submarine thriller *The Hunt for Red October*, which became a Hollywood blockbuster. It was about a Soviet breakthrough in submarine stealth technology. This was my territory.

****

Flag Officer Submarines' (FOSM) Headquarters had recently moved to the Joint Maritime Headquarters at Northwood in London, far removed from the sea but collocated with CTF345 and CTF311, the control cells for submarine operations, and also the RAF's Maritime Patrol Aircraft Command plus NATO's COMSUBEASTLANT (Commander Submarines Eastern Atlantic). It was also the headquarters of the Commander-in-Chief of the Fleet. Northwood was the epicentre of the country's Anti Submarine Warfare effort.

The Headquarters was an innocent enough looking place, tucked away in one of London's more expensive northern suburbs with a bus stop at the main gate, but appearances can be deceptive. Underneath its buildings was 'The Hole', a massive, bombproof, underground bunker that was the national nerve centre for our maritime operations. It was from here that our naval and amphibious wars would be fought. It was from here that Admiral Sir John Fieldhouse had managed the Falklands campaign, and it was from here that the Royal Navy and RAF would fight on through nuclear Armageddon. This would be the survival pod for some four hundred and twenty military personnel, four hundred being male and twenty being female – not the optimum ratio for regeneration of the human race. That was all very well but there was not enough living accommodation inside the fence and so I was billeted, along with two other Commanders, with a highly colourful landlady called Lulu.

Living with Lulu was a revelation. She was a statuesque, fifty-something, half-Mexican half-Italian woman of volatile temperament who lived apart from her sugar-daddy husband in a large, five-bedroomed house with three Commanders as paying guests – and an irritating little King Charles spaniel called Whisky. I find small dogs unattractive but Whisky had his merits. Lulu had two lovers, one her former driving instructor, the other a retired British Airways pilot who liked to remain naked during their weekend assignations.

When I returned one Monday morning from home, I found Lulu in a highly agitated state. As her lodgers had been away for the weekend, she had invited her pilot friend to stay and Whisky, never having seen a naked man before, had leapt up and bitten his willy.

On another weekend when I stayed over to finish some work, my computer crashed. It restarted without any trouble but shortly afterwards Lulu came barging into my room wearing only a transparent negligee. It was an intimidating sight and she was in a state of great distress.

'You did not come to save me,' she wailed.

'Save you from what?'

'I have just been electrocuted.'

'How was I to know that?' Now I knew why my computer had crashed.

Being a respectable married man, I was somewhat reluctant to enter her boudoir but as an electrical engineer, I felt morally bound to investigate the cause of her alleged electrocution. It did not take long. The live wire on her bedside lamp was touching the brass body – a death trap. She was lucky to be alive.

She threw herself on her bed and grasped my hand. 'Don't go,' she said. 'I'm shaking.' She was. She really had been electrocuted. I grimaced. If she had been killed, I would have been a prime suspect for murder. I was the only other person in the house. Now I was sitting on her bed holding her hand while she lay spread-eagled in her diaphanous nightie, modesty not in her mind. If her husband had walked in, he would never have accepted my plea of innocence. If not charged with attempted murder, I could have been cited in a divorce case. What would Kate have made of that?

****

Noise ranging of our submarines was now my bread and butter and from time to time I would be helicoptered to a remote, rocky island off the West Coast of Scotland to listen to our submarines strutting their stuff. The only building on the island was a small, prefabricated affair with wires coming into it from the sea. Inside, it was full of sophisticated digital sound analysers. Outside, the island belonged to the Great

Black-backed Gull, the world's largest gull, powerful, predatory and ferociously defensive of its territory.

Great Black-backs do not bother to build nests. They simply lay their eggs on rocks and defend them against all comers, including me. I was unaware of this until I stumbled inadvertently into one of their egg sites and was immediately attacked by the mating pair. They had wingspans of 1.7 metres (5 feet 7 inches) and were dive-bombing me, screaming past only inches above my head and trying to peck at my skull. I beat a hasty retreat.

Seagulls are large birds. I once picked up an injured Herring Gull near my home with the intention of taking it to a vet. Not only were its wings as long as my arms but from under my arm, its neck was long enough to attack my face.

At another secret noise range in a Scottish loch, I had planned a trial with *Valiant* who was manoeuvring up and down at periscope depth while Jack Revie, a twinkle-eyed Admiralty scientist of mature years, measured the distance offshore with a gadget made from a pair of Perspex strips and a wooden ruler in the best Barnes-Wallis tradition. I had wanted to assess the degree of surface effect when a submarine was propelling at periscope depth. In the flat calm waters of the loch, the result was immediately obvious. With masts down and the submarine propelling at slow speed, there was a perfectly visible wake. It was as if an invisible creature (or sea monster) was quietly moving up the loch.

That came as no surprise. It was what I had suspected. Having watched a Gannet glide on the uplift from our fin when I was in *Osiris*, I had good reason to expect a similar uplift in water. In calm water, if a submarine were to be at periscope depth for any length of time conducting, for example, a covert periscope reconnaissance of an enemy coastline, its wake would be clearly visible to a searching aircraft, even when its periscope was down. I don't think we had been sufficiently conscious of that.

I was never able to make the same assessment with a nuclear submarine powering along at high speed as that could not be done in the confines of a Scottish loch. It would obviously leave a significant signature on the surface. Unlike a surface ship, the wake of a dived submarine spreads in three dimensions. Satellites have been able to observe such wake effects.

Our submarines had now become so quiet that submarine-versus-submarine exercises had become virtually impracticable as neither submarine could find the other. The solution, for exercise purposes, was to augment their noise signatures with an underwater loudspeaker called a hydrosounder. The task fell to me and became an immediate headache because these commercially produced devices flooded as soon as the submarine dived.

In exasperation, I paid a visit to the manufacturer's factory where I was suitably impressed with the quality of their products. Their products were designed to play music in swimming pools and they had been unaware that submarines were using their kit. The hydrosounder designer informed me that he had actually given a demonstration to the scientists at AUWE of Beethoven being played underwater across Portland harbour.

'How did you feed Beethoven into your hydrosounder?' I asked.

'Very simple. I just plugged a cassette recorder into the amplifier.'

My head began to buzz. 'Could we do that in a submarine?'

'Of course. If you have the hydrosounder, you'll have the amplifier. All you need is a cassette recorder.'

My brain was now on fire. Submarines could play music to each other. The Brits could fire an Elgar and the Soviets could hit back with a Prokofiev! And why not? Did not Tchaikovsky defeat Napoleon at Moscow in 1812? More importantly, here lay the prospect of acoustic camouflage, which was something we had never considered. An underwater symphony orchestra gliding through the Barents Sea might raise the suspicions of the Soviet navy but the noise of a fishing boat would not. Could our submarines acoustically disguise themselves as fishing boats? And would there be any military benefit if they did?

My next thought was: could our submarines disguise themselves as Soviet submarines? Now that was worth a try. From Naval Intelligence, I obtained a recording of a Soviet submarine transmitting on active sonar and then arranged for one of our hydrosounder-fitted boats to run over the aforementioned noise range playing the Soviet tape. It worked. From the listening post ashore, it sounded exactly like a Soviet submarine pinging its way down the range. That was a breakthrough. In a dogfight, if a Soviet submarine were to hear its target transmitting on Soviet sonar, he would think he was in a friendly fire situation. And while he was pondering, we could nail him. The technicalities were far from perfect but I had demonstrated the feasibility.

Ideas were now flowing thick and fast. One of the great problems with submarines was how to communicate with them when deep, as radio waves do not travel far through the sea. For example, a submarine in the deep sweeper role ahead of a surface force could not communicate with it without coming up to periscope depth to transmit on radio. That took time and could both compromise its presence and lose the target. Nor could two deep submarines communicate with each other except at close range by underwater telephone. But with low frequency hydrosounders, I reckoned that such communication should be possible. That is what Sperm Whales do. They dive deepest of all whales and can go right down to the seabed to hunt for giant octopus. They also travel in family groups and, when diving to such great depths, need to keep in touch with each other. To do this, they transmit very low frequency tones (20 H z), which travel great distances in the sea. They are, I suspect, simply mooing like cows to let the others know where they are.

As my ideas attracted serious interest, I was allowed to set up a top-secret trial, code-named *Trial Tommo*, in which one submarine transmitted a very low frequency Morse code signal from its hydrosounder to another boat fitted with a towed array.

Hallelujah, the signal was received clear as day on the receiving boat's towed array signal analyser. It was another breakthrough. If the signal could be received on a submarine's towed array, it could also be received on a surface ship's towed array, or a seabed array, or be beamed up to an aircraft via a sonar buoy. I had demonstrated another new concept, but before I could take the idea any further, it was time to move to pastures new. That's life in a blue suit. My successor, however, tasked one of his technicians to develop my idea further and he did, but left the Navy to market it privately. The man had an entrepreneurial streak and set up his own company for the purpose. The intellectual property rights were mine but it was all in a good cause. When I met him, I wished him well.

****

One day, FOSM's Secretary came to the door of my office and said with a knowing wink: 'Who's a clever boy then?'

'OK, what do you want?'

'The Admiral wants to see you.'

'About what?'

'Don't know. He just said he wanted to talk to an electronic whizz kid.'

Now I knew he was bullshitting. I was certainly not an electronic whizz kid but trotted off obediently to FOSM's office.

'I was talking to Commander-in-Chief after the morning brief,' said Admiral Richard Heaslip when I entered his office. (He had taken over from Sandy Woodward). 'He was looking for a forward-thinking Engineer with bright ideas. So I volunteered you.'

My suspicions rocketed. C-in-C had an Engineer Admiral, several Engineer Captains, and a host of Engineer Commanders on his staff. Why would he have asked Flag Officer Submarines to find him a forward thinking Engineer? I could smell a very large rat but dutifully made an appointment to call on C-in-C, before putting together a folder of all my bright ideas. This could be my big moment.

At the appointed hour, C-in-C kept me waiting – he was a very busy man – and I had little more that five minutes of my allotted half-hour. He was utterly charming but not in the least interested in any bright ideas of mine – he had one of his own.

'Thank you so much for coming up,' he said, as if I had just climbed a thousand steps to his first-floor office. 'I wanted to talk to you about braking.'

'Breaking, Sir?' I answered enthusiastically. 'As in Citizen's Band radio?'

I was impressed. CB radio was all the rage at the time. This was clearly a very with-it Admiral.

'No, no,' he replied, 'aircraft braking.'

Now I really did smell a rat. Why would C-in-C, with a staff of professional Air Engineers at his command, have sought a submarine officer to talk about aircraft braking?

'I was landing at Heathrow the other day,' he added, 'and thought: what a waste of energy. Don't you think we could capture aircraft landing energy in a magnetic field?'

Small wonder he hadn't asked his own staff. I tried to imagine a jumbo jet with six hundred passengers on board flying into a mega electro-magnetic wind tunnel inside which it would hit invisible magnetic buffers; or be brought to an abrupt halt by a massive arrester wire stretched across the runway at Heathrow, as if on the deck of an aircraft carrier.

'Good idea, Sir,' I said, marshalling all my diplomatic skill. 'Magnetic fields have very short-range effects but it may be possible.' No, it would not. I pictured a Boeing 747 being flipped head-over-heels in a magnetic field because its polarity was the wrong way round.

'Excellent, I'd be most grateful if you would give the matter some thought.'

That was Admiral-speak for: 'Go away and find out.'

I sighed inwardly. The idea was a clear non-runner. 'Yes, Sir,' I replied. It was not the time for candour.

By chance, one of my Scout's parents in Helensburgh was the Professor of Aeronautical Engineering at Glasgow University and another was a British Airways pilot. So the three of us dignified C-in-C's idea with some serious thought.

My pilot friend gave me the weight and landing speed of a BAC 111 jet, a smallish passenger airliner, and by simple mathematics, I calculated the kinetic energy available for capture on landing. Theoretically, it was enough to boil water for ten thousand cups of tea. A good start, but with jumbo jets landing every few minutes at Heathrow, West London would soon be submerged in an ocean of tea.

My professor friend then pointed out that the big problem for aircraft is taxi-ing on the ground. A jet flying from Glasgow to London would use more fuel on the ground than it did in the air. I did some further calculations. If hydraulic pumps were fitted in the landing wheels, they could slow the aircraft down whilst pumping up pressure into a hydraulic reservoir. The process could then be reversed and the hydraulic pressure used to drive the wheels during taxi-ing, a reasonable idea as long as the braking effect did not cause the aircraft to do somersaults down the runway. In fact, London Transport was at that time experimenting with just such a system for London buses. Then I did some further sums and realised the ironmongery required would demand at least a dozen passenger seats. The idea was feasible but probably not cost-effective.

In a perfectly scripted Ministry style report, I forwarded my findings to Commander-in-Chief with a covering note drawing his attention to Etorre's Observation, something I had spotted pinned to a wall in the Admiralty Research Laboratory at Teddington. Etorre's Observation states that: *every man has an idea that doesn't work.* My report was duly returned with a tick and the word 'Noted' in red ink, the latter being the prerogative of Commanders-in-Chief.

****

It was time to move signature reduction to the top of everyone's agenda and not just in submarines. HMS *Sheffield* had been hit and destroyed by a French-made *Exocet* missile during the Falklands war because her radar reflecting area was too large. The gospel of signature reduction had to be preached to a wider audience and so, in collaboration with Commander Richard Thorne, a surface navy Mechanical Engineer and old friend from Manadon, we organised a Navy-wide Signature Reduction Conference in Plymouth. The Conference was a great success and, on completion, I set off to drive home to Helensburgh in my old Ford Cortina. It was June 18th, 1984.

As if on cue, the Cortina developed a worrying noise problem in its rear just as I left Plymouth. Above fifty-miles-an-hour, the noise was unbearable. If I were to have driven up the M5/M6 at seventy, I would have been deaf before reaching home, and if I kept below fifty, it would mean a fifteen-hour drive. So I stopped the car and jammed some Kleenex tissues hard into both ears. If I couldn't cure the problem, I'd treat the symptoms. By then, night was falling and the car heater no longer worked, so I also pulled on my towelling dressing gown, which happened to be at the top of my bag, and set off again.

Halfway between Taunton and Bristol, I was overtaken by a police car with blue lights flashing and an illuminated STOP sign in its rear window. I pulled on to the hard shoulder and stopped.

'Right, Sir, first you've only got one headlight,' said the roly-poly police sergeant in an amiable West Country accent, 'and that's just for starters. Second, your rear number plate light isn't working. And third, I can't read your number plate because it's covered in mud.'

'Ah,' I retorted, 'the headlight's just gone out.' It had been intermittent when I left Plymouth.

I climbed out of the car, walked round to the front and gave the dead headlight a thump. Nothing happened. I thumped it again with the same result. 'I guess the bulb's just popped,' I said, heading round to the back of the car where I took out my handkerchief, spat on it and began to wipe my number plate clean.

'Not like that,' said the sergeant. 'What you need is some grass.'

He went over to the verge, pulled up a handful of grass and began to demonstrate how to clean a number plate. 'And what do you do, Sir?' he asked, wiping away the mud.

'I'm in the Navy.'

'And what are you in the Navy? A sailor?'

'No, I'm an officer.'

He stopped wiping. 'And what rank would you be then?'

'I'm a Commander.'

At that, he stood up and looked at me in disbelief. 'You're a Commander! Well you ought to be ashamed of yourself. What will your sailors think when they see you driving a car like this into the dockyard?'

I had never given any thought to what my sailors might think about my car.

'Where are you bound for?'

'Helensburgh – up in Scotland.'

He looked me straight in the eye. 'Well, Sir, you just get that light fixed as soon as you get home.'

Later, I was approaching the bridge over the Manchester ship canal on the M6 when I saw my next police car. It was motoring slowly along in the inside lane and I was about to overtake it. I braked and waited for a large, articulated lorry to overtake and sneaked past on the blind side. The tactic worked but not for long. Within a minute or so, I saw the blue flashing lights in my rear view mirror and once more was on the hard shoulder being interviewed by the police, this time by Lancashire constabulary. They wanted to see my driving licence, insurance, and MOT certificate. I had the lot and that seemed to satisfy them.

'One of your headlights isn't working. Get it fixed as soon as you get home.' With that they let me go.

Two hours later, I was on the A74 just north of Gretna when, once again, I saw blue flashing lights in my rear view mirror. This time, the illuminated sign in the police car rear window read: FOLLOW ME. And so I was led into a layby. Now I would be dealing with the Scottish police and I feared they would be much less tolerant having seen them deal with rowdy football fans. First they circled my car, kicking each wheel in turn. Then they stared at the bonnet and banged on the dead headlight. Finally they came to the driver's window. They did not have a good bedside manner.

'Driving licence, insurance and MOT certificate,' one demanded. They were, I feel sure, not expecting to see these documents but I produced the lot. They checked them thoroughly and asked: 'Where are you going?'

'Helensburgh.'

'Well, fix that broken headlight as soon as you get there.'

With that I was released. I had just run the gauntlet of three British police forces with a defective car and all three had let me go in an entirely civilised manner. It was only then that I remembered I was wearing my dressing gown and had large wings of Kleenex tissue sticking out of each ear. The traffic cops must have been radioing ahead warning each other to keep a lookout for a nutcase driving a one-headlight Cortina with large lumps of paper tissue sticking out of his ears.

That night I was not the only one dealing with the police. There was a national miners' strike taking place and this was the night of the infamous Battle of Orgreave in which the National Union of Mineworkers had organised a mass picket to blockade and close the Orgreave coking plant in South Yorkshire. Some six thousand pickets had been bussed in from all over Britain's coalfields for the purpose. MI5, however, had infiltrated their organisation and a similar number of police officers from ten county forces, including fifty mounted police and fifty-eight police dogs, had been deployed to stop them. It was a ferocious punch-up with multiple arrests, after which the police were publicly vilified for their strong-arm tactics, which would result in half-a-million pounds being paid in compensation to injured miners.

There was great Cold War irony in these events. While the Royal Navy was preventing the onward march of Communism through Nuclear Deterrence on the high seas, the police were fighting an actual battle against Communism in mainland Britain. This was not normal industrial action. The National Union of Mineworkers, led by their Communist leader, Arthur Scargill, was intent on bringing the country to its knees by starving our power stations of coal. Scargill's aim was no less than a coup d'état. He wanted to topple Margaret Thatcher's Conservative Government. It was political warfare.

In the previous miners' strike in 1974, the Conservative Government of Prime Minister Edward Heath had been forced to declare a national three-day working week as coal stocks at the power stations dwindled. Heath had been forced to call a General Election and lost. This time the Government, under Mrs Thatcher, were ready for battle. It had built up sufficient coal stocks at the power stations to weather the strike.

I had great sympathy for the miners. They were fighting for their livelihood and for the survival of their industry. They were men who wanted to work and had a hellish way of earning a living, but their industry was dying. Foreign coal could now be imported more cheaply than British-mined coal. That said, I had equal sympathy for the police. They were merely doing their job of preserving law and order. That night, by chance, I had conducted a personal survey of three British police forces and every one had treated me entirely reasonably and with courtesy. The miners could have conducted a more peaceful, Gandhi-like non-violent protest, but the NUM leaders wanted a fight. Either way, the miners were bound to lose. The coal industry was doomed.

It was a sad time. One of the country's traditional industries, one that had powered Britain into being a world superpower during the Industrial Revolution, was on its deathbed. Nuclear power stations would provide clean energy for the future.

****

By the end of my time as Signature Reduction Officer, I felt that I had made a significant impact on the Flotilla. Ambiguity over responsibility for submarine signatures had now been resolved and placed with the Mechanical Engineers. My greatest claim to fame, however, was in my preaching. I had won the after-lunch slot at the annual Submarine Tactics and Equipment Conference. Toby Frere, my former saviour in *Andrew*, was now FOSM's Chief of Staff and tasked me with stopping everyone falling asleep after lunch. It was a blank cheque for me to put on a show and I had them rolling in the aisles whilst delivering some powerful messages on signature reduction. More importantly, everyone who mattered in the submarine warfare world (and beyond) was in the audience. I had made a name for myself. Fifteen minutes of being a comedian had won me far greater recognition than endless patrols as a submarine engineer. Out of the blue, it had given me the high profile that ambitious officers dream of and it would stand me in good stead for the future. As my mother used to say: '*Sell yourself, son. Nobody else will.*'

CHAPTER 21

Squadron Life

'Theirs not to reason why.
Theirs but to do or die.'

TENNYSON: CHARGE OF THE LIGHT BRIGADE

From Northwood, I moved to the staff of the Third Submarine Squadron, and life was now more relaxed. It was no longer about what I did but about what our flock of submarines did. I was to be a mother hen, though some of our chicks viewed Squadron staff as the enemy. It would, for example, be for the Squadron Captain to decide whether or not an errant Commanding Officer should be relieved of his duties, which happened only once in my time when the CO of one of our boats was arrested in Gibraltar for being drunk and the boat had to sail without him. He was entirely competent at sea but to be in command of a nuclear submarine calls for responsible behaviour at all times.

From time to time, an officer would be sacked for failing to meet required standards. The submarine's Captain would initiate such action but the relevant Squadron officer would be involved. If the officer concerned was of my specialisation, I would ride his boat for a few days to offer advice. In doing so, I remained ever conscious of my own humiliation back in *Otter*. There were virtually no bad officers in submarines, and very few who were incompetent, but some did suffer from lack of confidence or were on the wrong end of a personality clash, as I had been. I always did my best to rescue such officers whilst, at the same time, respecting the judgement of their Captains.

To sit in judgement over brother officers, Squadron staff had to have proven track records at sea, and we did. 'Johnnie' Cooke, the Squadron Captain, known as SM3, had joined the Navy with me and had commanded *Warspite* on Falklands patrols and the diesel submarine *Rorqual* before that. He was the son of an Admiral, and set the standard for relaxed sophistication. Commander Tim Lightoller, the jovial but tough Squadron Executive Officer, had commanded *Turbulent*, one of our latest nuclear boats. His grandfather had been an officer in the ill-fated *Titanic*. The somewhat eccentric Commander Tony Quade, Squadron Mechanical Engineer, had been Senior Engineer in *Resolution* and had served in *Dreadnought*, Britain's first nuclear-powered submarine. We all knew our stuff.

In well-led submarines we would be welcomed on board but in others we were treated with deep suspicion. The Executive Officer of one boat, who later became an Admiral, always insisted that ship's officers dined with Squadron riders as honoured guests. In another, where the Executive Officer was not promoted, we were shunned and left to dine alone. It's so bleeding obvious. I found it astonishing that fellow officers could treat visiting Squadron staff as anything less than submarine royalty. After all, we processed their promotion reports.

Personnel reports are the lifeblood of the Navy. In these, officers are assessed against ten characteristics and marked out of ten in each as follows:

0–3	Non homo sapiens
4	Special reporting required with a view to dismissal
5	Average but unfit for submarines
6	Above average but taken as average in submarines
7	Doing well
8	Walks on water
9	Author lacks credibility
10	Non terrestrial origin

This meant that to be credible promotion candidates, officers had to score 75–78 points. Differentiation within such a narrow band called in turn for suitably persuasive pen pictures. I developed a talent for interpreting cryptic messages, such as: *has common sense beyond his years* (aged thirty-six); *has the potential to inhibit success*; *recently a father for the third time and a committed family man*; *an athletic young officer who invariably lands on his feet*; *his legal training assures he is fleet of foot*.

To interpret such bewildering comments, Squadron officers had to be familiar with the personalities, and so we went to sea to observe our flock in action. There could be, for example, a personality clash between a Commanding Officer and his Senior Engineer. If the latter happened to be in a promotion zone, an unsympathetic Captain could easily destroy his career. The reverse also held true as sympathetic Captains could exaggerate the merit of their favourites. Thus, it fell to Squadron officers to form a balanced judgement across all boats.

I saw my role as that of a coach in that I had to choose the first eleven but also encourage the reserves. Combining honesty with encouragement, however, is difficult when you know that a good, hard-working officer is not suitable for higher rank in which different skills are required. After all, promotion-winning football coaches often have to replace old-faithfuls with new players better equipped for the higher league.

At the bottom of an Officer's Report was a small box for the Squadron Captain to make his recommendation for promotion, with about twenty-six words allowed. Here the cryptology moved up a level: 'Not Yet' meant 'Never'; 'Recommended' meant 'Not Yet'; 'Promote Now' meant 'Recommended'; 'My top candidate' meant that this officer was in pole position against the best candidates from other squadrons. I fear that many good officers would have joyfully accepted that they had been 'recommended' for promotion when in reality, they had little chance. The subsequent disappointment could break a man's heart.

Dealing with young sailors was an entirely different experience. A seventeen-year-old Mechanic called Johnson, straight out of training, was attached to the Squadron as a messenger whilst waiting to join his first submarine. Although I was not a coffee drinker, the young man kept bringing me cups of coffee. When I asked him why, he replied: 'Commander Quade told me that my only purpose in life was to keep the coffee coming.'

When I reminded Tony Quade, who owned Johnson, that a young trainee will take his orders literally, he went ashen. 'Shit!' he groaned. 'I've just told him to take my in-tray out of my sight and set fire to it.'

Some months later, I was boarding one of our submarines at sea and there, standing on the casing resplendent in a rubber diving suit, was MEM Johnson.

'Hello, MEM Johnson,' I said as if greeting an old friend, 'are you training to be a diver as well as a submariner?'

'Oh no, Sir,' he replied. 'I can't swim.'

'Can't swim? Then why are you dressed in a diving suit?'

'Cos I'm swimmer of the watch, Sir.'

'How on earth can you be swimmer of the watch if you can't swim?'

'Because I'm a Senior Rates' messman.'

'But you can't swim.'

'That's all right, Sir. The other messman can't swim either and he's swimmer of the other watch.'

'So what will you do if you have to jump over the side to save someone?'

'The Cox'n told me that if I have to go into the water, I've to crack open this air bottle and my suit will inflate and keep me afloat.'

This young sailor stood ready to turn himself into a Michelin man and throw himself at the mercy of the sea in the service of his country. As Tennyson said of the Charge of the Light Brigade: *Theirs not to reason why. Theirs but to do or die.*

Leading Mechanic Holt was an entirely different kettle of fish. He was an Edinburgh hard man, an adult-entry and, I suspect, had already begun his criminal career before joining the Navy. He listed 'nightclub bouncer' as his previous occupation. Holt had been landed from one of our submarines to face criminal charges for having broken into the Vendopac machine in the NAAFI arcade and was under stoppage of leave pending trial. Like Johnson, he had been attached to the Squadron office as a temporary messenger.

On learning of his previous career, I asked him if he would be willing to act as a bouncer for my Scout discos, which were being plagued by older, alcohol-consuming boys trying to gain entry. He agreed willingly and I arranged for him to be granted temporary leave under my personal supervision. As a bouncer, he was simply superb. He was no taller than I but whereas I wore a heavy sweater and quilted anorak to stave off the cold at the Scout Hall door, he stood in an open-necked, short-sleeved shirt and dared any of the local ruffians to come anywhere near him. And when they threatened to call in their big brothers as reinforcements, he just said: 'Bring them on.' This young man had natural power of command. He was smart, polite, and could not have been more helpful.

At that time, Lieutenant Phil Hudson, my deputy, had decided to join the Royal Australian Navy and I had planned that SM3 would present him with a Squadron crest as a farewell present. The officer-in-charge of the Base workshops, however, declared that his budget was spent and refused to manufacture one. As the costs involved were trivial, I was damned if I would to take no for an answer and so asked Holt to pop down to the workshops and persuade a chum to make me one unofficially. The tactic worked to perfection. A couple of hours later, Holt returned clutching a Squadron crest. It had two little holes through it which had evidently been filled-in and touched up with fresh paint, but was otherwise perfect.

That evening, Phil Hudson was duly presented with his Squadron crest by SM3 and the following morning left for Australia. That same morning, the unflappable Johnnie Cooke summoned the entire Squadron staff to his office. It was the first time I had seen him angry. 'Someone,' he announced incandescently, 'has just stolen the Squadron crest from the front of my staff car! – And the workshops are refusing to make me a replacement.' I kept a straight face. He had presented Phil with the crest from his own staff car.

I once conducted a qualifying board for a young Mechanic who had been cross-trained as an electrical switchboard operator. The young man had given perfect answers to all my questions until it came to the electrical safety question. 'What action would you take,' I asked, 'if you found someone lying unconscious with his hand in the switchboard?'

He remained silent.

'Come on,' I said. 'A man with his hand stuck in the switchboard, obviously being electrocuted, don't you think you could save him if you killed the volts?'

He thought for a moment and then, with great authority, said: 'Oh no, Sir. If you only kill the volts, the amps will still get him.' (!)

He had learned all the switchboard drills by heart and I had no doubt that he would have manned his switchboard through hell, fire and brimstone but he had no concept of electrical theory. In his mind, electrical cables were the conduits for two murderous tribes: the Amps and the Volts. So what? Few of us understand our car engine but still drive the car.

Training young men (and now women) with limited educational qualifications to operate enormously complex equipment safely and reliably is one of the greatest triumphs of the modern Navy.

****

Amidst the plethora of routine signals on my log, one would occasionally raise a smile. *Valiant* sent this PRIORITY signal just after departing for a two-month patrol in the Falklands. It had to be serious. It was:

From: HMS VALIANT
To: SM3

Chief Petty Officer Scroggie has failed to provide for family pet rabbit called Warren during current patrol and wife and children have gone south. Request Naval Patrol provide food and water until July.

In similar vein, correspondence in my in-tray could raise a smile.

Sunshine Inn
South West 46th Street
Fort Lauderdale
Florida

Her Majesty the Queen
Buckingham Palace
London

    Dear Queen
    I'd like to send my warmest and gracious thank you on having the pleasure of meeting and serving your Royal Navy, the Boys in Blue on the HMS CHURCHILL, during their stay at our hotel, the Sunshine Inn.
    My manager, Buck Saddlesore, and myself and the rest of our crew in the restaurant were so impressed by their pleasant attitude and politeness that it not only impressed us but the whole hotel, from the maids to the front desk clerks, to the pool keepers, even our other guests that were staying here. They never got out of hand and had such a beautiful time. God's blessings over the Boys in Blue on HMS CHURCHILL wherever their tasks may take them.
    You should be very proud because we sure are.
    Sincerely yours,
    Emmylou Gadomski

The reply from Buckingham Palace read:

Dear Emmylou,
I am commanded by Her Majesty the Queen to reply to your recent letter about the
visit of HMS CHURCHILL to your country. Her Majesty was pleased to learn of the
impression that members of the Royal Navy had on you and your colleagues. It is always
gratifying when people take the trouble to express their thanks in writing.

****

To ensure that our submarines were up to standard, Squadron staff would ride each
boat for its pre-patrol Index period in which the complete repertoire of submarine
drills and emergencies would be exercised in a simulated war scenario. Our task
was to initiate emergencies, observe, assess, and where necessary, correct the boat's
response. To help in this we would borrow specialists from the Captain Submarine
Sea Training's (CSST) staff, their main role being to train new crews in boats coming
out of build or major refit. They were not family like the Squadron – they were
tough, uncompromising trainers who turned new crews into efficient fighting units.
I felt suitably honoured therefore when Warrant Officer John Andrews from CSST
took me aside during *Sceptre's* Index.

'We don't normally allow officers to do this,' he said confidentially, 'but we'd
like you to start the Short War for us with this thunderflash.' He handed me a
thunderflash. 'Sneak up to the Torpedo Compartment and at eleven o'clock precisely,
drop it into one of the gash cans. And for Christ's sake, don't forget to leave the
lid off, otherwise it will be fired into the air.' (A thunderflash has similar explosive
power to a hand grenade). 'You do know how to trigger a thunderflash, don't you?'

'Of course,' I replied, feeling the dead weight of responsibility bearing down on
me. It felt as if I were about to bat for the entire officer corps.

The scenario was that the boat had suffered enemy attack. My explosion would
trigger a carefully timed sequence of consequential dramas: the tripping of main
generators, loss of electrical supplies, flooding in the engine room, fire in the galley,
Emergency Stations, Action Stations, and the setting up of a counter-attack. All
hell would be let loose and all would have to be dealt with in breathing apparatus
because of the resultant fire, smoke canisters being used for realism.

As *Sceptre* was a new class of submarine for me, I snuck forward surreptitiously
to identify a suitable gash can and found one just off the top of the ladder in the
Junior Rates recreation space. At the appointed time, I climbed the ladder again,
lifted the lid off the gash can, triggered the thunderflash, dropped it in, left the lid
on the deck, and beat a hasty retreat. A few seconds later, there was an almighty bang
at the forward end of the passageway followed by the sounding of the submarine's
General Alarm and the cry of 'Emergency Stations.' I felt very pleased with myself.

I had discharged my thunderflash in the right place at the right time and with the desired effect.

At the end of the exercise, I happened to pass the Leading Seaman responsible for cleanliness in the Junior Rates recreation space. 'You wouldn't believe it, Sir,' he said. 'Some fucking idiot let off a thunderflash in a full gash can and our deckhead (ceiling) is covered in shit!'

****

For her war patrol in the South Atlantic, *Valiant* had been equipped with a trial version of the American *Sub Harpoon* anti-ship missile, which was about to be introduced into RN service. As she was due for major refit on return and the kit would become redundant, it occurred to me that it could be transferred to *Odin*, the last remaining diesel boat in the Squadron, as she too was scheduled for a Falklands patrol. A brilliant idea, I thought, but the Ministry of Defence Procurement Executive vetoed it. *Sub Harpoon* for diesel submarines was not in their costings. In retaliation, I set up the Alternative Procurement Executive, to be known by the acronym APE, and organised a Sub Harpoon Initial Trials programme, to be known as APESHIT. The Ministry was not amused.

As I knew little about *Sub Harpoon*, I then devised a new management technique called 'Management by Objection'. The principle was extremely simple.

Me: 'I intend to fit *Sub Harpoon* in *Odin*.

Objector: 'You can't do that.

Me: 'Why not?'

Objector: 'Because you need a left-handed flip-flop in the Forends.'

Me to Lt Dolton: 'Can you fit a left-handed flip-flop in your Forends?

Lt Dolton: 'Yes.'

And so on. Objectors thus mapped out every step required. Far from putting obstructions in my way, their objections were stepping-stones to success. The last thing I needed was someone who said: 'Oh what a brilliant idea.' That was no help at all.

Success followed success and *Odin* (Lieutenant Commander Malcolm Avery) duly deployed to the South Atlantic with *Sub Harpoon* missiles. She was and remains Britain's only missile-firing diesel boat. I was hugely proud of this achievement but could feel the pins going into my effigy in Whitehall.

****

Some of our boats were engaged in spying and when one moves into the realms of intelligence gathering, strict need-to-know rules apply. Those who need to know have to sign classified registers clearing them for knowledge of a code word, sometimes only to cover a small part of an operation. Once given code word clearance, I could then discuss matters with other signatories on the list. Records of signatories were kept in an office inside a large safe – the X-registry. Any discussions on the subject had then to be held inside an electronically screened bank vault called the Secure Conference Room.

I found myself signing up for one code word that gained me entry into an obscure workshop in the Base of which I had no previous knowledge, and to which entry was permitted only at night. It was like a submarine version of Q's workshop in the James Bond movies. My task was to organise the movement of a very large tarpaulin-covered object from the workshop to the jetty to fit into an about-to-depart sneaky boat under cover of darkness. The operation required the closing of the Base main road and a police escort. Under the tarpaulin was a massive, hydraulically operated claw. When it had been fitted to the submarine, she looked like a gigantic, one-armed lobster. Before dawn, she had sailed. I was not cleared to know about her operation – I did not need to know that – but I suspect her mission was to snatch some Soviet listening device for intelligence assessment.

Intelligence gathering is one thing but disinformation is another. For amusement, I had a photograph taken of my Alsatian dog wearing earphones in a sonar operator's training booth. When asked by the Base Photographic Officer why it had been required, I replied tongue-in-cheek: 'We're planning to replace sonar operators with dogs because they can easily be trained to recognise the sound of Soviet submarines and bark if they hear one, and dogs are much cheaper than sailors.'

Somehow this was leaked to the press and appeared in Navy News along with my photograph. I've often wondered what Soviet Military Intelligence made of that.

Life in the Squadron was fun.

CHAPTER 22

The Directorate of Naval Lost Property

'Bureaucracy is a giant mechanism operated by pygmies.'

HONORE DE BALZAC

With the Conservatives standing at a healthy fifty per cent in the opinion polls, Margaret Thatcher went to the country and I went to my Appointer. She won a third term in office (1987) and I won a plum appointment to the Directorate of Naval Operational Requirements in the Ministry. I was to produce the Staff Requirement for our next class of nuclear submarine, code-named SSNOZ, an absolute dream appointment. At last, my talent had been recognised, but it wasn't to last. A few weeks later, the appointment was vetoed by my Director-to-be who thought I would cause trouble because of the Marconi incident with my *Thompson Torpedo* and for fitting *Sub Harpoon* in *Odin*. It was outrageously below-the-belt.

I was then appointed to DNLP, which I assumed to be the Directorate of Naval Lost Property but was actually the Directorate of Naval Logistic Planning. As any half-qualified warmonger will tell you: *when it comes to war, amateurs talk tactics, professionals talk logistics.* DNLP was the Planning Directorate for Fleet Support, one of the largest organisations in the entire Defence establishment. The Chief of Fleet Support managed the Royal Dockyards, the Royal Fleet Auxiliary (tankers, stores and ammunition ships), naval aircraft, armament depots, the flotilla of marine auxiliary craft such as tugs, aerial farms, telephone exchanges, and the entire supply chain that provided victuals, spare parts, transport, clothing and medical supplies. It was very big business indeed.

DNLP was based in the massive, green-roofed, white Portland stone, Ministry of Defence Main Building in Whitehall, the epicentre of British defence, within shouting distance of Downing Street and missile range from Moscow. Here sat the Secretary of State for Defence, the Chief of the Defence Staff, the First Sea Lord, the Chief of the Air Staff, the Chief of the Army General Staff, the Principal Under Secretary for Defence (the top Civil Servant), and now me.

As my office was a garret on the top (ninth) floor, I had literally gone up in the world. I could gaze down on Horse Guards Parade, hear regimental bands playing at the Cenotaph and the clip-clop of horses pulling carriages to Buckingham Palace with official State visitors. As I sat in my eyrie, I felt more like Christopher Robin than a steely-eyed submariner.

The Main Building stood on the site of the Tudor's old Whitehall Palace, the wine cellars of which now provided succour to the warriors of the twentieth century. Here, Henry VIII had married both Ann Boleyn and Jane Seymour. Here Shakespeare's *The Tempest* was first performed and here, in front of the adjacent Banqueting Hall, Charles I was beheaded. Ominously, the old palace was destroyed by fire in 1698, a diarist of the time noting that: 'Whitehall burnt! Nothing but walls and ruins left.' If the Cold War went hot, history would repeat itself.

In these grandiose surroundings, I was received with supreme indifference, except for the office coffee boat for which I was given immediate responsibility. I was to share the garret with three other officers of my own rank, in Ministry terms the lowest rank of any significance. One was a Fleet Air Arm pilot who was primarily occupied in setting up a pyramid-selling company in the City. The second was an Engineer of dubious sanity from the surface Navy who having drunk himself into pie-eyed oblivion one lunchtime, returned the following morning convinced that he had been arrested in Charing Cross underground station and taken away for interrogation by MI5 or the KGB – he wasn't sure which. The third was a roly-poly, bald-headed, divorced Civil Servant in his early sixties called Bob. I never did discover what he did. His main interest was a dating agency that was helping him to find new love in his miserable life. The agency did the trick. One morning he stepped off Cloud Nine and came skipping into the office announcing that he had just met the perfect woman and was head-over-heels in love. I was extremely happy for him.

My new boss, the Director General of Fleet Support Policy and Services (DGFSP&S), known as 'Fish and Chips', was none other than Admiral Toby Frere in whose shadow I had been walking since he first rescued me in *Andrew* almost twenty years earlier. As he knew me very well, I had no need to make a good first impression but neither could I afford to fail him. His was the policy-making department behind the whole massive Support operation and I was to be responsible for Corporate Planning and Manpower.

The task was challenging. For a start, I failed to discover what my predecessor had been doing. He seemed simply to have been swimming against the current in an endless torrent of itinerant Ministry files. Nor could I see any distant objective on which to focus. This was not the frantic, target-focussed, operational pressure cooker of deploying submarines to sea on time. Life in the Ministry was a merry-go-round shrouded in impenetrable administrative fog and revolving around the annual ten-year budget setting exercise known as the Long Term Costing (LTC). The challenge was to smuggle in your pet project without it being detected, like when the Labour Government cancelled the Navy's aircraft-carrier programme, only for it to creep back in as a 'through-deck cruiser'. The original atomic weapons programme was so thoroughly well hidden that it did not appear in the Estimates at all. As real-time operations such as managing depots and dockyards were farmed out to rusticated Director Generals, most of the work in the Main Building focussed on the ifs-and-maybes of the future.

I was now at the heart of a massive bureaucracy. Those at the top kow-towed to our political leadership in Westminster. Those at the bottom simply kept shovelling paperwork from their In to their Out trays, or into their Pending trays if indecision was the preferred option. My task seemed to be to generate enough paperwork to keep them shovelling, but on what subject? That was the big question. Where did I begin?

I consulted Julian Malec, a close friend from Manadon and now an established Whitehall warrior. 'It's very easy,' he said. 'Find out who else is doing your job and copy him.' A nearby office door bore the message: *If the boss calls, take his name.*

As my In-tray stood tall and was regularly topped up by clerks, the thrill of landing a file in my Out-tray felt like scoring a goal but I had to speed up. I had to learn the Civil Service technique of quickly annotating files with meaningless comments dressed up to sound positive, such as (actual quotes): *'I recognise that your novel proposals are a constructive way to move the debate forward'; 'the propriety of this interpretation does give some cause for concern'; 'given the difference in views, I could not claim that a consensus exists and have therefore forwarded a holding reply which does not foreclose any options.'*

Being non-committal was a Ministry virtue but I was a simple soul. I dealt in black or white, go or no-go. My idea of management is: 'There is the goal. Here is the ball. Now score.' In the Ministry, this translated as: 'The goal is debatable. We can't afford a ball. Now write a discussion paper.' I would have to up my game in the creative non-fiction stakes.

Whilst still pondering my role, I was sent to the prestigious Cranfield School of Management for a short course on corporate planning. On arrival, I was psychometrically tested and scored the highest possible mark for entrepreneurialism. Small wonder I felt like a fish out of water in the Ministry – I was in the wrong game.

One of the big Defence issues of the time was whether or not women should serve at sea. Politicians were strongly in favour but the Navy Board was dragging its heels. It was proposing that women should be allowed to go to sea but only in Fleet Auxiliaries and survey vessels, the Board's main argument being that women would not be strong enough to handle heavy stores during transfers at sea or cope with the physical demands of damage control in battle. These were flimsy arguments and the Admirals needed to find a way of getting the politicians off their backs.

Having worked closely with Wrens in my torpedo days, I was well aware that women were a great untapped pool of talent. Therefore I submitted the proposal that women should go to sea in the new Trident submarines that were about to enter service. Deterrent submarines were, by definition, non-combatant, never transferred stores at sea, and did not have to face damage control scenarios. Trident boats would be ideal for female crew. They even had four-berth cabins, which would allow separate accommodation without the need for expensive alterations. It was the perfect solution. It would put women into the Navy's highest priority programme and thus get the Navy Board off the political hook. The submarine community disagreed. In an unofficial straw pole, the submariners voted to admit homosexuals rather than women, both being forbidden at the time.

In the end, Archie Hamilton MP, Minister for the Armed Forces, pulled the rug from under us by decreeing that women were to go to sea in all combat-ships except submarines. Once again, I was ahead of my time (by twenty-six years); in 2014, women were finally permitted to go to sea in Trident submarines.

When discussing the matter with the Commandant WRNS, I played Devil's advocate. She was pushing hard for equal opportunities for the girls and thought that they should go to sea in submarines.

'There's no way women can go to sea in nuclear submarines,' I declared provocatively.

'Why not?' she replied, rearing up.

'Because they would have to be given pregnancy tests immediately before sailing.'

'Why?'

'Because radiation would have serious consequences for a developing foetus.'

My tongue was largely in my cheek. Outside the Reactor Compartment, radiation levels in a submarine are lower than when sunbathing and much lower than flying in an aeroplane. When a submarine dives, the ambient radiation level actually drops because the sea shields it from the sun. My Devil's advocacy held sway for twenty-six years.

****

Whilst beavering away at my corporate plan, the world as I had known it turned on its head. The Cold War came to an abrupt end with the fall of the Berlin Wall and within two years, the Soviet Union had dissolved. Our enemy had packed it in.

There was now to be a Peace Dividend. Translated, that meant massive cuts to Defence spending, a complete re-think on our Long Term Costing, and a deluge of additional work for Ministry warriors. There would now be redundancies in the Armed Forces. Career prospects would evaporate. And did we still need a Strategic Nuclear Deterrent?

What brought about this political earthquake was not military victory but the combination of economic pressure on the Soviet Union and the arrival in 1985 of the fifty-four year old reformist Mikhail Gorbachev. He had succeeded the geriatric

Communist dinosaurs, Andropov and Chernenko, who had died in quick succession. Gorbachev had inherited massive economic problems, an escalating Cold War arms race, and was faced with President Reagan, who was upping the stakes on American defence spending, including the 'Star Wars' Strategic Defence Initiative. Gorbachev knew that things could not continue in the old Cold War way now that the Soviet economy could no longer compete. He wanted *perestroika* (rebuilding) and *glasnost* (openness).

But could he be trusted? It had taken Gorbachev weeks to admit that there had been a major nuclear disaster at Chernobyl, and that had raised questions over his true commitment to *glasnost*. However, President Reagan and Prime Minister Thatcher believed in him and so, incredibly, the Cold War was ended through good interpersonal relationships. Gorbachev had charm and exuded sincerity. To quote Mrs Thatcher: '*I like Mr Gorbachev. We can do business together.*'

Arms reduction treaties were now agreed and former Warsaw Pact Soviet republics gained independence from Moscow, most opting for democracy, and many aspiring to membership of both NATO and the European Union. If this caused a planning upheaval in Whitehall, one shudders to think of the mayhem in Moscow. It was the greatest world event since the end of the Second World War. For the first time in my life, I was not at war with an enemy that had the power to destroy me.

In 1990, Mikhail Gorbachev was awarded the Nobel Prize for Peace – *Time* magazine named him Man of the Decade and ex-President Nixon called him Man of the Century. Sadly for Gorbachev, his popularity at home did not match his popularity in the West. Six years later he was removed from power by the very democratic system he had enabled.

Always expect the unexpected. In August that year (1990), Saddam Hussein's Iraqi army invaded Kuwait and we were at war once again. That was a revelation. The Ministry went to Action Stations! The corridors of power suddenly buzzed with urgency, as if on the front line. Meetings were called to show action was being taken. Directors took to sleeping on camp beds in their offices. The lids of the Treasury coffers were prized open. Money was no object. If we needed something, we could have it. All we had to do was slap 'Urgent Operational Requirement' on the request and it was granted. It was as if Santa Claus had arrived in the Ministry.

****

One night Kate phoned to say that she had noticed a property advertised in the local paper that seemed to match what we had in mind for our final resting place. I viewed it that weekend. It was a pair of simple, conjoined hillside cottages facing west with a panoramic view of the upper Clyde estuary. To cut a long story short, we made the maximum bid we could afford and our offer was accepted. No financial risk was involved as a sale had been agreed on our bungalow and the top-up mortgage for the new cottages was within our affordability.

That March, my father died and I ceased to be F G Thompson Junior. I managed to fly up from London and arrived at his bedside just before he passed away. He was so intensely proud of my career in the Navy that I chose to believe the rumours in Whitehall and whispered in his ear that I was about to be promoted to Captain. I don't know if he heard me – he was too far-gone to show any response – but I like to think that he met his Maker happy with that knowledge.

The rumours proved correct. That December I was promoted to Captain. At this late stage, my career was accelerating. But then the wheels came off. The sale of our bungalow fell through and there were no other offers. I had to take out an expensive bridging loan to cover the new mortgage and then discovered that the cottages had dry rot throughout and would have to be completely gutted at substantial further cost. To make matters worse, the Thatcher Government had recently introduced poll tax in Scotland. This penalised second-home owners to the tune of double poll tax. Legally, I now had two empty cottages plus our existing home, so I was due to pay five times the normal rate. I could not afford that. To rub salt into this gaping financial wound, my salary was also about to drop. Engineer Captains were not entitled to Submarine Pay and that was greater than my pay rise. In one small leap for mankind, I was facing bankruptcy.

I had never been in debt before. I had always lived within my income. I had taken calculated risks but never with money. I was beside myself with anxiety. It was the sort of crisis that could drive a man to a nervous breakdown, or alcohol, or even blow a marriage apart, but I was not in danger of any of these. Submarines had taught me to grit my teeth and carry on, and I had Kate as my firm foundation. She never complained but in this ruinous financial situation, I felt that I had failed her. Paralysed with anxiety, I packed my bags and headed north to take up my new appointment as Captain Submarine Maintenance (Chief Engineer) at Faslane.

CHAPTER 23

The Nuclear Business

'If you are going to sin, sin against God, not the bureaucracy.
God will forgive you but the bureaucracy won't.'

ADMIRAL HYMAN J RICKOVER USN

In my new appointment it was back to nuclear duty. I was cast in the dual roles of Captain Submarine Maintenance and Chief Staff Officer (Nuclear), the former having responsibility for keeping the active submarines operational, and the latter for the assurance of nuclear safety in what was the largest nuclear reactor maintenance operation in Europe. I was officially schizophrenic. I could red card myself. I was also a member of the Base Board with an engineering staff of six hundred and sixty men and an annual budget of £24 million. My organisation could undertake all technical work in submarines except for nuclear refuelling. I felt thoroughly competent to handle these responsibilities and was driven by the all-consuming imperative to ensure that there would not be a nuclear disaster on my patch. I would rule my roost with a rod of iron.

The Navy simply could not afford to have a nuclear accident. All other issues apart, it would bring our entire nuclear submarine programme crashing to a halt and could even threaten its very existence. In my new duties, I was directly responsible to Commodore Tom Blackburn RN, one of the most polished officers I have met. He was a fast tracker, two years younger than I, immaculately groomed, hair parted with mathematical precision, and with a vocabulary utterly devoid of vulgarity. So smooth was he that after his time as an Admiral, he was appointed to Buckingham Palace as Master of the Household to the Sovereign, serving Her Majesty the Queen directly. The last time I saw him, he was in top hat and morning coat as part of the Royal entourage at a garden party in the Palace of Holyroodhouse in Edinburgh, but he greeted me as an old friend. To quote Kipling: *If you can talk with crowds and keep your virtue, or walk with Kings – nor lose the common touch.* That was Tom. He seemed so superior yet was completely unpretentious. But appearances can mask tragedy.

Tom was not a submariner. He was the son of a submariner. His father had been Captain of HMS *Affray* when she sank in the English Channel in 1951 with the loss of all hands, including an entire class of Sub-Lieutenants (Seaman and Engineers) from HMS *Dolphin* who were embarked only for a week of training. The loss of *Affray* has never been fully explained but material failure whilst snorting is assumed to have been the cause.

The nuclear industry is the only industry to have begun life with safety as its first priority, at least in the West. Unlike most other industries, safety consciousness was built-in from Day One and the safety disciplines at Faslane were already impressive. Not a spanner could be lifted on nuclear work without a formal written procedure, approved by the Procedure Authorisation Group (PAG), nor could any man enter a radiation environment without strict Health Physics control.

The PAG was a multi-lateral committee consisting of representatives from all parties involved including Rolls Royce and Associates, the reactor design authority, each member having the right of veto. Every step in every procedure had to be unanimously approved and every operative had to sign for every step he had completed. When the task was finished, the signed document was returned to the PAG for audit before the work was verified as complete. It was tedious and meticulous Quality Assurance and should have left no scope for mistakes.

In addition to nuclear safety, rigorous regimes were also required for submarine and human safety. We did not want a repeat of the *Warspite* fire nor any fatal accidents nor a sinking like that which happened to the old diesel submarine *Artemis* in 1971. She had just finished a maintenance period and was being refuelled when sea level reached her after hatch. As that had been left open, the sea came flooding in but the hatch could not be shut because heavy electrical supply cables had been led through it. The boat then sank rapidly alongside the jetty in *Dolphin*, the traditional home of the Submarine Service. Fortunately, no lives were lost but three men were trapped in the forward torpedo compartment and had to escape in immersion suits through the escape tower. The incident was a huge embarrassment and a salutary reminder that minor carelessness can cause major accidents.

Sadly, safety standards in the rump of the old diesel submarine fraternity had slipped behind those in the nuclear boats and so the Submarine Sea Training organisation was established with Captain Sam Fry, a veteran submarine CO, put in charge with the remit to get tough. He would work up all newly formed crews to the highest of standards and re-test them from time to time in conjunction with their parent squadrons. Standards would not be allowed to slip again.

Submarine maintenance work is replete with hazardous activities, each having the potential to go seriously wrong. Ultimately, it was my responsibility to ensure that accidents did not happen.

Hazards can come in surprising forms. Two years after women were integrated in the Navy (1993), Royal Naval Temporary Memorandum 64/95 was issued: *In the past two years there have been numerous serious injuries, including six amputations, caused by personnel catching rings or jewellery on objects such as hatches or machinery. Personnel are reminded of the danger of wearing jewellery on board Her Majesty's Ships* – unless, of course, you are Her Majesty.

In view of this litany of potential disasters and the number of nuclear submarines undergoing maintenance work, it seems almost miraculous that after fifty years of operations, there has never been a major accident in the Clyde Submarine Base. This is the result of ultra-professionalism and rigorous safety management. The reality is that virtually all accidents can be avoided by simple safety-first procedures. Had her after hatch been shut, *Artemis* would not have sunk.

It is carelessness, poor management, and negligence that cause accidents, or irresponsible pressure from senior management that forces mistakes. The latter caused the Chernobyl disaster, the fatal rail crash at Hatfield, and the tragic sinking of the *Herald of Free Enterprise* with the loss of one hundred and ninety-three lives. None of these tragedies could have happened under the safety management regime at Faslane.

In the case of Chernobyl, engineers had been forced to override safety systems to meet a deadline. At Hatfield, engineers were not allowed to close a railway line to replace known cracked rails because of commercial pressure from the train operating companies. The *Herald of Free Enterprise* sailed with her bow doors open to meet her sailing schedule and sank before she had even left Zeebrugge harbour. That was equivalent to a submarine diving with its conning tower open. Incredibly, the Captain had sailed without even knowing if his bow doors were open or shut – there was no such indicator on the bridge. Such utter disregard for basic safety measures is incomprehensible to me as a submariner. I shudder to think of the ramifications were a nuclear-powered submarine to sink in harbour and kill one hundred and ninety-three innocent civilians due to such negligence and corporate irresponsibility. We were thoroughly switched-on to safety, but there was never room for complacency.

Whilst accidents can be avoided, material failure is a different issue. Things can and do break, but even so there are always reasons. To avoid accidents due to material failure, we had rigorous non-destructive examination programmes in place. These involved pressure testing, X-raying or ultrasonic scanning of all critical items such as hull valves, hull welds, and reactor system welds. Cracks in any one of which could have had serious consequences, as illustrated so tragically in the *Thresher* disaster. Here, our submarine design authorities knew what and when to test. A small crack is not in itself dangerous – one can drive safely with a cracked windscreen – but cracks tend to propagate and that can lead to catastrophic failure. The trick was to spot a crack in its infancy and take corrective action.

As a result of this regime, cracking was identified in *Warspite* in a transition weld between dissimilar metals, where cracking is more likely to occur than between similar metals. The crack discovered was inside a steam generator where stainless steel pipes of the reactor's primary circuit were joined to the non stainless steel metal of the boiler. The problem was known as Trouserlegs because the two pipes involved looked exactly like that. Were this crack to have propagated, it could have led to a primary coolant leak with serious consequences. That set a cat among the Design Authority pigeons. Was this peculiar to *Warspite* or a generic fault? We had to know. All submarines had to be checked at the earliest possible opportunity and any similar cracks repaired immediately. It was a gigantic headache for everyone concerned including me. The welds in question were virtually inaccessible without completely dismantling the submarine. Worse still, they were in a relatively high radiation environment. Only a custom-built robot could undertake the work.

This sudden demand for inspection and repair in all our nuclear submarines led quickly to a logjam at the jetty. At its peak, I counted seven submarines alongside, all with their reactor compartments opened up and nuclear inspection work in hand. They were, in effect, grounded. Rolls Royce engineers rose to the challenge impressively by designing an ultra hi-tech robotic system that had to be assembled inside every steam generator. It was like building a ship in a bottle or repairing an artificial heart in a live patient using keyhole surgery. My nuclear repair team had to carry out all the preparatory work and the buttoning up afterwards whilst my nuclear safety watchdogs had to follow every step of every procedure. It was the biggest technical challenge in the history of the nuclear submarine flotilla and the storm broke whilst I had responsibility for nuclear safety.

Manpower then became the problem. Radiation workers such as stainless steel welders were rare enough commodities and they were limited to a permissible annual radiation dose. When that had been reached, they had to be laid off on to other work for the rest of the year. Thus, we had to import highly skilled men from national resources and that brought the additional challenge of integrating strangers into our management system. Our nuclear repair teams were plunged into round-the-clock, seven-day-week shift working with personnel sometimes having to switch between boats and jobs in mid-shift, a perfect recipe for mistakes. (On one unrelated occasion, a Petty Officer from Base Staff went on board one submarine to conduct routine maintenance but chose the wrong boat – they all look the same from the outside – and by the time the poor man had finished his task, he was at sea).

Although the Cold War had ended, the game was not over. Russia still had massive nuclear-armed forces and Russian SSNs were still trying to locate our SSBNs on patrol. Two American SSNs actually suffered collisions with Soviet boats in the Barents Sea in 92/93. Our top national priority, however, remained continuation of Deterrent patrolling. As we had only two Polaris boats in the operational cycle – one out, one in – the Trouserlegs problem caused their patrol lengths to be increased to an eye-watering one hundred and twelve days in one case whilst work was completed in the other boat. For one hundred and fifty men to remain locked up in a steel tube for one third of a year without going mad, making mistakes, or even complaining (or being paid extra), testifies to the quality of our nuclear submariners.

As the Ministry of Defence was already planning to cut submarine numbers, Trouserlegs accelerated the decommissioning of the ageing *Conqueror*, *Warspite*, *Churchill* and *Courageous*, all being scrapped in 1990/91 with *Valiant* following three years later. But the *Resolution* Class SSBNs, which were of similar age and with many more miles under their belts because of their two-crew system, had to continue operating.

My own role in all this was as piggy-in-the-middle. I had to keep the pressure on to complete the work and return the boats to service whilst also ensuring that no mistakes were made. That did keep me awake at night.

The volume and pressure of nuclear work was one thing but it was entirely dependent on people and the people-factor must never be overlooked. It was inevitable that sooner or later a mistake would be made. When it was, no harm was done, but it set my alarm bells ringing.

We used liquid nitrogen to create freeze seals when there was no isolating valve to shut off a pressurised pipe. That meant building a small bath around the stainless steel pipe involved and filling it with liquid nitrogen to freeze the water in the pipe. The ice then formed a plug as good as any stop valve and the pipe could be cut. It was a very simple and effective technique, but liquid nitrogen boils at minus 196 degrees Celsius and is continuously boiling off, so the bath had to be topped up regularly.

For this reason, the procedure specified a liquid nitrogen sentry, his sole duty being to keep topping up the bath from a large flagon. It was just like pouring water, but if you were foolish enough to stick your hand in it, you would lose it. It would be like sticking it into a fire except this was extreme cold. For the poor sentry, usually a young, semi-skilled mechanic, it was an utterly boring job.

On this occasion, one of my nuclear welders had been hanging upside down in a tiny space inside a hot sticky reactor compartment welding back together a cut pipe. Out of sympathy for the young sentry, he had given him a break and undertaken the task of topping-up the nitrogen bath himself. Predictably, he became so pre-occupied with his welding that he forgot about the bath and the ice plug began to melt. Had it done so, it would have led to a leak of reactor coolant water. Fortunately, he spotted his error in time, took remedial action and was honest enough to admit the mistake, confession being a standard nuclear discipline.

On hearing of this, I went ballistic. It was just the sort of human mistake I had feared. I stopped all reactor work immediately and declared that it would not re-start until I was satisfied that no further mistakes would be made. We were doing too much. We were overloaded. Our key men were becoming tired. The shift working was relentless and most of the tasks required difficult work in unpleasant conditions. There were immediate howls of protest from Submarine HQ: how dare I stop such high priority work? Pressure was heaped on me to call off the ban but I had played my red card and no one could override it. I was the referee.

I then ordered the welder concerned to be brought before me. I would rip him up for arse paper, demote him, hang him from the yardarm *pour encourager les autres*, and kick him out of this and any other navy.

When the poor man was marched into my office, I took a deep breath. He was my best and most experienced welder, the very backbone of my nuclear repair team. He was a Chief Petty Officer with thirty years of unblemished service in the Royal Navy and it was blindingly obvious that being hauled before his Captain as an offender was more than punishment enough. He needed no reminding of his mistake and would never make it again. So I changed my mind. I looked at the schedule of his recent work, congratulated him on his outstanding efforts, his impeccable record as a nuclear welder, and on his long and loyal service to the Royal Navy. When he left my office, I sat down and wrote a commendation for the award of an MBE. It was duly awarded. That too would encourage the others. Not many welders make the New Year's Honours list.

A less satisfactory outcome was reached with Commander Z, my deputy. I had to sack him for attempting a technical shortcut. It was the first time I had sacked anyone. I had acquired the nuclear ruthlessness required by Admiral Rickover.

In dealing with people, one can rarely predict what comes next. My secretary, a shy, courteous, personable, young Lieutenant in his first job was the same age as my elder son. He was very willing but gauche beyond belief and caused me a number of embarrassments but I felt sympathetic; I had struggled in my own first job. I could cope with most of his shortfalls but when he locked himself out of our joint office, lost his keys, borrowed mine and then lost them as well, that did test my patience.

When the phone rang at home one Sunday afternoon, my adrenaline system fired into action. It could only mean there was a crisis in the Base but no. It was my young secretary. He was phoning to wish me a happy birthday. It was the only time in my thirty-seven years in the Navy that anyone had done that – in my macho world, it simply wasn't done.

He also kept bringing me cups of coffee despite my telling him that I preferred tea. To solve the problem tactfully, I presented him with a box of teabags. Fifteen minutes later, he appeared in my office holding a tea bag as it if it were contaminated.

'What do I do with this, Sir?' he asked, head cocked to one side like an inquisitive budgie.

'You make tea with it!'

I began to wonder about the young man's origins. It was as if he had been brought up in a particularly reclusive monastery. In the end, I concluded that I was a bad role model for him. I was far too non-conformist in handling correspondence, simply refusing to let an in-tray rule my life. He needed to work under an experienced secretary for proper professional development. So I had him transferred to the secretarial staff in Fleet headquarters.

Some months later, I saw his name associated with an officer who was being dismissed from the Navy for being homosexual. I don't know if my young secretary was gay nor do I know how I would have handled it had he come to me for advice on the matter; homosexuality was then a dismissal offence in the Armed Forces. My only possible counsel would have been to keep it secret or resign before being dismissed in disgrace. I fear that acute anxiety over his sexuality may have been the cause of his seemingly permanent state of pre-occupation. He must have felt trapped. In retrospect, I find it tragic that he felt unable to seek my advice without risking dismissal. Gay, lesbian and transgender personnel were finally allowed to serve in HM Forces eight years later (2000).

Although my production department was called the Naval Technical Department and had originally been all-Navy, it now included Civil Servants, both industrial and non-industrial, as a *quid pro quo* for the insertion of uniformed personnel into civilian departments following the Civil Service strikes at the time of my leaving *Revenge*. The policy was called Mixed Manning.

This brought into sharp contrast the differences in personnel management, pay, and conditions between the Navy and the Civil Service. For example, Civil Service industrial workers were paid much less than their uniformed equivalents and were therefore happy to extend jobs into overtime, particularly at weekends, when double or even triple time was paid. Naval personnel wanted jobs finished as quickly as possible as that was the Naval instinct.

Similarly, a Naval technician could go to the Senior Rates Mess at lunchtime and have a pint of beer. A Civil Service technician sharing the same workshop would be sacked for consuming alcohol at work. Predictably, this led to some highly sensitive clashes in discipline. I would happily have banned alcohol altogether but the Base was home for the uniformed staff who lived on board. For the civilians, it was merely their place of employment.

The other big difference was that in the Navy, officers are responsible for the welfare of their men (and now women). If a man had a personal problem, he took it to his Divisional Officer. It is a system as old as the Navy itself and all Naval Officers have the instinct to care for and speak on behalf of their men. In the Civil Service, personnel matters are dealt with more impersonally, via a personnel department or via trade union representatives.

In my career to date, I had never had a civilian lady as an employee but had recently recruited a thirty-something, middle-ranking, female Civil Servant to draft a business plan for my department. She became one of only three or four women amongst my six hundred and sixty men and would come to see me from time to time with draft management plans. One morning, she popped into my office looking timid as a mouse and placed an envelope on my desk.

'What's this?' I asked.

'It's my resignation.'

I was shocked. 'Have I upset you?' I asked, searching my soul for what I had done to offend.

'No.'

'So what's prompted this?'

She blushed. 'My marriage is breaking down. I need more time at home.'

'I'm very sorry to hear that,' I replied, genuinely saddened. She seemed such a nice lady. Had she been a sailor, I would have said: 'Sit down and let's talk about it,' but I did not want to be accused of intruding on her private life. I had no responsibility for that.

'It's not for me to comment on your private affairs,' I added, 'and I'm not qualified as a marriage guidance counsellor but it seems to me that if your marriage is likely to break down, the last thing you should be doing is resigning from your career in the Civil Service. You have a good job and a secure salary and with cuts in the Civil Service, once out, you may not get back in again. I know only one divorced woman and the most vital thing for her was economic independence.'

We never discussed the matter again but she withdrew her resignation.

Some time later, I found an envelope on my desk with a card from her. It read: *Thank you for everything.*

'Thanks for the card,' I said, next time I met her, 'but I haven't done anything.'

'Yes, you have,' she replied. 'You listened.'

Listening to the problems of others costs nothing yet can be of such value.

****

The Base was accustomed to an endless procession of VIP visitors but our visitor on 4th December 1991 was someone completely different. With the Cold War over, relationships with Russia had thawed and we were being visited by Army General V N Lobov, Chief of the General Staff of the Soviet Armed Forces and First Deputy to the Russian Minister of Defence. He was our former enemy's top military commander and was visiting *Revenge*, my old boat, whose missiles had once been targeted (possibly still were) on him and his colleagues. Afterwards, I had lunch with him and as a memento of his visit, presented him with a beautiful pewter model of a Polaris submarine mounted on a varnished mahogany plinth, manufactured in my workshops. In response, his aide pulled out of a cloth bag the tackiest imaginable Perspex model of a Soviet frigate and presented it to me. It looked as if a first-year apprentice in some Soviet training establishment had made it. (I presented it to the craftsman who had made Lobov's gift).

The fascinating thing about Lobov was that he was not in the least interested in Polaris or Trident or any of our nuclear submarines. He probably already knew all he needed to know about them. His only interest was in our pay. He had worked out that as head of the Soviet armed forces, he was being paid the same as one of my Petty Officers! I was tempted to explain my own desperate financial situation but didn't know if Communists did mortgages.

Happily, my personal financial crisis was now reaching its end. My bungalow had been sold and the bridging loan had been paid off. My mortgage had now come down to the manageable proportions originally envisaged, but I had no capital left for renovating my derelict cottages. So we shut up the smaller one and Kate and I moved into the other to live like squatters. To make matters worse, Scottish Power cut off our electricity pending our supply cables being relocated underground and for good measure, the Local Authority condemned our water supply, taken from the nearby burn, as unfit for human consumption. It was a bizarre situation. I was returning from heavy managerial responsibilities and hobnobbing with international leaders to digging trenches for electricity cables whilst teaching myself how to build a water sterilising station that would meet Local Authority drinking standards.

****

The elephant on my horizon was Trident. The Base was about to welcome HMS *Vanguard*, the first of our new Trident submarines. The project had begun in the dark days of the Cold War (1982) and would provide a massive increase in our nuclear retaliation capability but *Vanguard* was arriving two years after the Cold War had ended and that threw an avalanche of fresh fat on to the political fire. The difference between Trident and Polaris was only quantitative but *Vanguard*'s arrival rejuvenated the dwindling, anti-nuclear 'peace camp' that had been established at our main gate in 1982.

Everything about Trident was vastly greater than Polaris. A Polaris submarine weighed eight thousand tons, while a Trident boat weighed twice as much. A Polaris submarine could fit inside a Trident one. The Trident 2 missile had a range of twelve thousand kilometres, three times greater than a Polaris missile, and each carried a Multiple Independently Targetable Re-entry Vehicle (MIRV) with a capacity for up to twelve nuclear warheads. The missiles could also remain on board for the whole of a submarine's commission (the period between major refits).

Vanguard was simply too big for existing Base facilities and so, in a twist of historic irony, a new Trident base was being constructed in the old shipbreaking yard next door where the previous *Vanguard*, the battleship I had dreamed of commanding as a small boy, was broken up for scrap. Work on the new base had begun in 1985 and included a state-of-the-art jetty with all the facilities needed for the new class, plus a massive shiplift capable of raising a sixteen-thousand-ton submarine out of the water. When complete, I would inherit responsibility for the new facilities.

For the past twelve years, the Trident Project had been the country's top Defence Procurement priority. It had also carried the nation's credibility in the eyes of the Americans. Britain was not simply buying Trident. In opting for the Trident 2 system, Prime Minister Thatcher had bought into the American development programme as a 5% shareholder. That meant that British companies were involved in the research, development, and manufacturing process. It also meant that Britain would be sharing missiles with the US Navy in what was called 'mingled stock' i.e. when cycling missiles for maintenance, British submarines could be loading missiles that had previously been in an American submarine and vice versa. As a country, we had to match American standards and programme dates. However, while the missile system was American, the submarines, nuclear warheads, and support facilities were all British. The Chief Strategic Systems Executive (CSSE) in London, Admiral Ian Pirnie, a submarine Electrical Officer, was managing the project.

As a future stakeholder, I had been attending all relevant project meetings and knew that the new jetty would not be ready in time. It was a serious programming issue and I had voiced my concern but was ignored. We were facing the arrival of *Vanguard* and would have nowhere to berth her. As no one else seemed willing, I would have to blow the whistle. By my calculations, six weeks would be needed to convert two of our existing jetties to accommodate *Vanguard* and that I set as my deadline for Plan B.

By coincidence, Tom Blackburn was about to hand over to John Trewby as Commodore and, I suspected, would not wish to toss in this managerial hand grenade as his parting shot. Equally, I guessed that the new Commodore would not wish to begin his tenure by challenging the omnipotence of CSSE's project management, but my deadline had been reached. The changeover of Commodores was my opportunity to strike. While the two Commodores were engaged in their handover, I seized the day and sent a matter-of-fact signal to CSSE and all other Trident-related parties marked: *From Captain Submarine Maintenance*. It said simply: *As new Trident jetty will not be ready in time, intend converting five and six berths for Vanguard arrival.* Boy, did the shit hit the fan.

The following morning on John Trewby's first day in office, Admiral Pirnie phoned him in a rage. I had to be emasculated, disemboweled, and then burned at the stake. Who the hell did I think I was? Never again was I to be allowed to send such a signal. I had caused major embarrassment in Whitehall. Thus, in my first interview with the new Commodore, I was given his most frosty 'we are not amused' look and received the required chastisement. I could sense that he was thinking: how do I rid myself of this troublesome officer? I must have seemed exactly the sort of chap to ruin his career.

Yet again, I had been branded as a naive troublemaker but I was neither naive nor making gratuitous trouble. Nor did I repent. Mine had been a carefully calculated intervention. I had played an ace and Admiral Pirnie would have known that. He could not deny the facts. What would have infuriated him was that someone from outside his all-powerful project had forced the issue. I was thoroughly proud of my intervention. To delay the inevitable any longer would simply have created a last minute crisis.

Vanguard duly arrived on time and berthed without a hitch at our newly modified jetties. A major project milestone had been achieved on schedule. A crowd of thousands gathered on the promenade in Helensburgh to watch her cream past and head up through Rhu Narrows into the Gareloch where I was waiting on 'my' jetty to greet her. She was awe inspiring, truly enormous. There was much rejoicing and self-congratulation but not a whisper of recognition for my critical role in the success. I didn't care. I knew I had saved the programme.

****

John Trewby, the new Commodore, was another fast tracker but not a submariner. He too had joined the Navy two years behind me, had read Engineering at Cambridge, and was well on his way to becoming an Admiral. And he came with an impeccable Naval pedigree with both his father and grandfather having been Admirals. John's father, also an Engineer, had gained an international reputation for introducing gas turbine propulsion to warships and had been responsible for completing the original Polaris programme before promotion to the Navy Board as Chief of Fleet Support. Despite our difficult start, John and I rapidly formed the most cordial of working relationships and have remained friends ever since – but he did have his revenge.

He received an invitation from the Duke of Argyll to join a shooting party and keen not to offend His Grace, accepted. He duly returned clutching a set of antlers, which he passed to me with a request that my workshops mount them on a wooden shield. As I am opposed to killing animals for pleasure and to spending taxpayers' money on having the fruits of such labour mounted, I put the antlers on top of my cupboard and forgot about them.

Antlers in the hands, or rather head, of a stag are lethal weapons, as I discovered the following morning. When I opened my cupboard, they fell on top of me and I was gored by a dead stag. They could easily have pierced my cranium and damaged my precious brain.

****

The aristocracy were a relatively familiar sight at the Base. In July 1992, the All Party Parliamentary Defence Study Group visited, including the Earl Attlee, Viscount Mersey, Lords Craig of Radley, Astor of Hever, Judd, Kennet, Marlesford, and Vivian, and the delightful Baroness Jean Cherry Drummond of Megginch (created 1628). Alas, when the Commodore was leading this august ensemble on a tour of the floating dock, they were showered in diesel oil from *Repulse*. As the dock was my department, I had then to issue the following letter to the Captains of the submarine squadrons:

Gentlemen,

The Commodore was acutely embarrassed by a diesel carnival that occurred in the floating dock during the visit of the All Party Defence Study Group. As the noble lords and ladies take less pleasure from showering in diesel than we submariners, would you please ensure that in future, submarines in dock withhold all excretions until guests have departed?

Veering from the sublime to the ridiculous is implicit in submarine life. One unexpected consequence of *Vanguard's* arrival was an accusatory letter from the Clyde River Purification Authority, which had been struggling to explain an increase in raw sewage washing up on the foreshore at Helensburgh. This it linked erroneously to the arrival of the larger Trident submarines. What was the Navy going to do about it? The letter was passed to me for action. I replied as follows:

Dear Doctor Shanks

1. *Thank your for your letter of 25th October on the subject of sewage discharge from Trident submarines.*

2. *Though physically much larger than their Polaris predecessors, Trident submarines have reduced crews and it is the crew, not the submarine, which generates the biological waste. As Trident submarines are equipped with a shore discharge facility, you may be assured that they make no contribution to the contents of the Clyde.*

3. *You will be pleased to note that from 1995, all submarine sewage will, like its Commanding Officer, be piped ashore.*

****

When the phone rings at six o'clock on a Monday morning, it is rarely good news. 'Prepare for a rough ride when you come in,' said the Base duty officer. 'We went to Shelter Stations last night. It was a false alarm but was reported to the press. The Scottish dailies are carrying the headline: *NUKE DISASTER AT SUB BASE.'*

A submarine had been warming up its reactor prior to sailing and was discharging excess primary coolant into a portable effluent tank, a routine procedure. Water expands when heated and shrinks when cooled. When a reactor is shut down, extra water is pumped in and then drawn off again when it warms up. On this occasion, the hose connection had leaked and a jet of hot water had squirted on to the submarine's casing. On witnessing this, a Ministry of Defence policeman had pressed the nuclear alarm button – there was one on every jetty – and that triggered a full nuclear emergency with sirens going off and Base personnel rushing to Shelter

Stations. Once a nuclear alarm has been pressed, the emergency drill cannot be stopped. A false alarm had then to be declared but the peace-campers at the main gate had heard the sirens and contacted the media. When I arrived, a BBC team was waiting to interview me.

'How radioactive was the spill?' I asked the Base Environmental Safety Officer in preparation for the interview.

'About as radioactive as a cup of coffee.'

'About as radioactive as a cup of coffee,' I answered when the BBC interviewer asked the same question – a pretty good sound bite I thought. At least it was until the British Coffee Association threatened to sue me for defamation of their product.

The fact is that natural radioactivity is everywhere. One kilogram of coffee has about 1,000 Becquerels. We are born with it in our bodies; an adult human has about 7,000 Becquerels. Brazil nuts have it, granite buildings in Aberdeen have it, Cornish houses have it, hospitals and luminous watches have it, but when it was associated with the Base, it was always reported as a disaster. The anti-nuclear campaigners made sure of that. As I had a team of radiation monitoring personnel constantly mapping the radioactivity profile of the surrounding area, I could prove whether or not a Base event had any radiological implications for the local populace, unlike the 'peace campers'.

The leak should not have happened. It was a reportable Incident but was trivial.

'Incident' and 'Accident' are definitive terms in the nuclear industry: an Accident is *an unplanned release of radioactive contamination into the atmosphere* whereas an Incident is merely an unintended event. Although I never expected to be dealing with a real Accident, it was nevertheless one of my top three priorities for I was also responsible for our Nuclear Accident Response Organisation (NARO). In this, I was no longer hidden away in the nuclear world but had to deal with public organisations such as the police, local and national government departments, the media, and politicians.

Despite never having had an Accident, the Base was under constant media attack by highly motivated anti-nuclear activists, not least from the so-called 'peace camp' on our doorstep. A small corps of professional anti-nuclear consultants had also evolved and become convincing voices when nuclear matters were in the news. They were expert at turning everything into doomsday scenarios such as: *most of Scotland will be affected (CND)*. Their credibility was further enhanced by the Ministry's habit of issuing bland statements such as: *the probability of an Accident is more remote than 1 in 200,000 years of reactor operations* – from such statistics, Joe Citizen could reasonably conclude that this could be the year. CND's response to that statement was: *The risk of an accident in a nuclear powered submarine is much higher than for a shore-based reactor.* It is certainly true that nuclear submarines present far more demanding reactor operating profiles than civilian power stations but submarine operators actually live with their reactors, a compelling reason for staying safe.

In assessing risk, probability analysis is used and professional mathematicians produce the Ministry's statistics. The claims of anti-nuclear groups, on the other hand, were of more dubious validity. After Chernobyl, for example, the World Health Organisation estimated that up to 4,000 delayed deaths had occurred due to radiation effects while Greenpeace put it at 200,000. Who was right? As Benjamin Disraeli once observed: *there are lies, damned lies and statistics.* Thus, I found myself embroiled in a public relations battle, not a real nuclear issue.

It was not an easy battle. We were constantly on the defensive. Activists could say what they liked, could attack when they fancied and would go to the media without the full facts, always putting the Navy on the back foot. Publicity was their aim; avoiding it was ours. Ambitious left-wing politicians would arrive for demonstrations at the main gate simply to gain nationwide publicity from being arrested – the only time they were to be seen there. No one demonstrated in our favour.

Our press releases had to be cleared by the Ministry press office in Whitehall – Crown servants must never embarrass a Minister – therefore our responses tended to be slow and often obscure: *can neither confirm nor deny.* Nor could I eclipse the anti-nuclear rhetoric with positive news such as: *forty years without a nuclear accident at Faslane.* In a democracy, *ALL'S WELL* does not sell newspapers.

When the adjective 'nuclear' is introduced, objectivity goes overboard and fear of the unknown kicks in. In this respect, the anti-nuclear propagandists have been hugely successful. They have turned 'nuclear' into a toxic word and the media stands ready to sensationalise anything vaguely nuclear. Philosophically, this is not an entirely bad thing. It is democracy in action. It's called being held to account. In non-democracies, the authorities control the media and hide the truth, like at Chernobyl.

The nuclear industry is certainly not infallible – no human endeavour is – but its failures should be viewed in perspective. In the sixty-one years between 1952 and 2013, I count only 110 nuclear Accidents globally, most being trivial or localised e.g. mishandling of a radiotherapy source. Only three have been serious enough to include reactor core damage, namely Three Mile Island in the United States (1979), Chernobyl in the USSR (Ukraine) (1986) and Fukushima in Japan (2011). Of these, only Chernobyl and Fukushima reached the maximum score on the international scale for nuclear Accidents and both could have been easily avoided. In Chernobyl, operators disconnected safety control systems to speed things up. At Fukushima, a higher sea wall or waterproofed electrical supplies would have prevented the Accident altogether – it was loss of electrical power to the cooling pumps that caused the problem.

Such Accidents were undeniably disastrous but the fatality count tells a less dramatic story. There were no fatalities or adverse health effects at Three Mile Island, except for cardiac arrests due to stress in the traffic jams caused when 140,000 people voluntarily evacuated the area following an uninformed debate on local radio. The debate had speculated over what might happen rather than reporting what was happening. Thus it unleashed the fear factor. Evacuation was unnecessary.

At Fukushima, according to a United Nations report, there were no deaths or serious injuries due to direct radiation exposure, although six workers died from non-nuclear causes. The report estimates that one hundred others may have suffered late cancer deaths whereas six hundred died due to evacuation. These relatively small casualty figures should be set in the context of over 15,000 deaths caused by the actual tsunami. Chernobyl was a completely different story. That was an utter disaster in every respect; even so, only fifty or so direct fatalities occurred.

By my count, only sixty-one nuclear-related fatalities have been recorded in the lifespan of the global nuclear industry, including Chenobyl. It is virtually impossible to know the numbers of consequential cancer-related or incidental deaths but if one takes a blind guess at 50,000, this should be set in perspective against the World Health Organisation's figure of 1.25 million road traffic deaths worldwide in a single year (2015), or the unknown millions of smoking related cancer deaths, or the 3,787 deaths in India caused by a release of methyl isocyanate gas from the Union Carbide pesticide plant at Bhopal where 558,125 members of the public were injured. Automobiles also pump out atmosphere-changing greenhouse gases yet there is no popular demand for an end to the automobile industry. Why not? – Because automobiles have societal benefit. And so has nuclear generated electricity.

Why is there such antipathy towards the nuclear industry when it is one of the world's safest industries and is not a greenhouse-gas-emitting source of power? It is because of the success of anti-nuclear propaganda, media sensationalism, and the failure of pro-nuclear public relations. I find that very frustrating.

The public safety plan for the Base was produced in association with the Civic authorities and maintained by my department. It was available in local libraries though few members of the public were aware of it. It was not a paper tiger. Every year, a major three-day exercise called *Short Sermon* took place, rotating annually round the Naval nuclear operating bases. These exercises involved both naval and civic authorities and were played right up to national government level, the Minister for the Armed Forces always participating in person.

These exercises started with an imaginary Incident in a submarine that developed into an Accident leading in turn to the worst-case scenario of reactor core meltdown. The complete disaster had to be simulated to allow all agencies to be tested, for example in the processing of irradiated casualties in local hospitals unused to handling such cases. Public relations were at the core of these exercises as effective communication with the public is crucial in any large-scale calamity to prevent blind panic and to provide vital information on such matters as whether or not to evacuate. In these matters the police have primacy. But producing timely press releases in a fast-moving emergency is not easy. First there is the inevitable fog of uncertainty in establishing the facts and then the need to establish consensus with all participating authorities on actions required. Over-reaction can lead to panic and under-reaction can lead to casualties. Indecision leaves the field wide open to media speculation and the airing of unauthorised opinion by self-styled experts as happened at Three Mile Island.

Exercise press conferences were therefore like show trials, an array of professional and student journalists being briefed to attack spokesmen as aggressively as possible, including the Minister who would fly up from London for the grilling. Dr John Reid MP, Labour Minister for the Armed Forces, hugely impressed me. He arrived from London at the height of one exercise, had an intense thirty-minute technical briefing, took to the stage and handled every question like a seasoned nuclear expert.

When the exercise was not at Faslane, I acted as a senior peer-group assessor in the other Naval nuclear areas.

****

After four years on the engineering frontline, I was appointed to the staff of the Chief Strategic Systems Officer (CSSE) in Whitehall as Assistant Director of Programmes and Safety on the Trident Project, there being no more jobs at Faslane for an Engineer of my rank. My career was no longer under development. I had reached the stage where the Navy had to deploy my experience to best advantage.

CHAPTER 24

The Chief Strategic Systems Executive

'The atomic bomb made the prospect of future war unendurable. It has led us up those last few steps to the mountain pass; and beyond there is a different country.'

J ROBERT OPPENHEIMER. LEADER OF THE MANHATTAN PROJECT

I felt a buzz on returning to Whitehall but came down to earth on entering the Main Building. I had forgotten what a soulless place it was. At least I had gone up in the world. I had descended from the Ninth to the Second floor, which in Ministry terms was closer to the seat of power.

CSSE was now Rear Admiral Richard Irwin, a fellow Engineer who had been one year ahead of me at Manadon and had also been in diesel boats before specialising in the Polaris Missile System. We knew each other well and enjoyed mutual respect. Though completely down to earth, his naturally superior style was enhanced by the wearing of a monocle and occasionally the wearing of a cloak and the carrying of a cane.

The Trident Project had just reached its high water mark in public and political profile and House of Commons committees, in particular the Public Accounts and Defence Select Committees, were frequently calling Richard to Westminster for interrogation, journalists noting every word for later dissection. This was democracy in action and not the forum for a faux pas. I would help with his brief and occasionally accompanied him. Fortunately, CSSE was in a position of strength, the project having delivered spectacularly well. Whether or not one agrees with Trident, as a feat of project management it was a huge national success story. Its climax came shortly after I joined when *Vanguard* successfully fired her test missiles on the range at Cape Canaveral. For me, it felt like the party was over and I had arrived for the tidying up.

To acquaint myself with the project, I flew out to King's Bay in Georgia, the main base for the American Atlantic-based Trident submarines. It covered sixteen thousand acres, was home to ten Trident submarines, several other submarine squadrons, and had swamps full of alligators. The Americans certainly think big. They have a Can Do attitude and can afford it. The British have to rely on Make Do but we were matching their standards in Trident. The King's Bay Trident development had begun in 1976, with ours in 1985. Their first Trident boat had deployed in 1989 and *Vanguard* deployed four years later. We were not far behind and they treated us with the greatest of respect. They had no need for us but seemed to like having us on board – probably for the entertainment value. This was the legendary Special Relationship in action. We were true partners-for-peace and enjoyed a uniquely strong bond of trust.

From King's Bay, I flew to Washington to meet my opposite number, the softly spoken, bespectacled Captain Walt Elliot USN, the most laconic of laconic Americans. He had just completed a tour in command of a 'boomer', as SSBNs are called in the US Navy. We had never met but took to each other like blood brothers.

'You must have Scottish ancestry,' I said. 'The Elliots were Border reivers.'

'Yeah, I got married in Iona when I was based at the Holy Loch.'

'Did you know that your official reception tomorrow night happens to be on the anniversary of the birth of Robert Burns, Scotland's national bard?'

'Gee no, Eric,' he replied with the animation of a rampant sea slug. 'Could you write a speech for Admiral Nanos to give?'

Admiral Nanos was Head of what the Americans called their Special Projects Division. He was an utterly charming Greek-American on his way to the highest ranks of the US Navy. I doubted if he would have heard of Robert Burns.

'Afraid not, Walt,' I replied. 'I don't have any reference material with me.'

The following afternoon, on a visit to Arlington Cemetery to pay our respects to America's war dead, Walt arrived clutching a large leather-bound book. 'I got this for you from the Arlington library,' he said. 'It's a copy of *The Complete Works of Robert Burns*.'

My face bounced off the edge of President Kennedy's grave. I had three hours to write a speech for an American Admiral I had never met on a subject about which I knew little, and it was to be delivered at a formal US reception.

Back in my hotel room, the book fell open at a well-thumbed page. It contained the poem, *Ode For General George Washington's Birthday*. I had never heard of it but knew that Burns had been a great supporter of the American War of Independence, viewing the rebels as fighting for the just cause of freedom. Burns thoroughly embraced the American Declaration of Independence: *We hold these truths to be self evident: that all men are created equal; that they are endowed by their creator with certain inalienable rights; that among these are life, liberty and the pursuit of happiness.* It was the basis for Burns' own anthems for humanity, *Scots Wha Hae*, the greatest call to the defence of freedom ever written, and *A Man's a Man*, the greatest evocation of man's equality with his fellow men.

Sometimes I surprise myself. The hidden strengths one finds when cornered are unpredictable and can be impressive. In three hours, through a fog of stress, I crafted a stimulating speech embracing the literary genius of Burns, the US-UK Special Relationship, the Trident missile system, the merits of British project management, and finishing with the stirring line: *What unites us this evening is our common cause, the defence of freedom.*

After giving the speech, Admiral Nanos was to propose the toast: 'Her Majesty the Queen.' Burns would not have approved of that. He refused to drink toasts to 'The King.' (George the Third of the madness fame in his day). He would propose an alternative toast such as: 'The health of a better man, George Washington.' All very well but I could not propose that Admiral Nanos replace his toast to 'Her Majesty the Queen' with 'The health of a better person, Bill Clinton.'

At the reception when I was almost blinded by migraine, Walt came over and played his joker. 'Admiral Nanos has been delayed in the Pentagon. He would like you to give his speech.' I thought my head would explode.

In the event, I was saved by the bell; we had waited so long for the Admiral to appear that there was no time for my speech.

The business of our joint talks was mundane, covering such detailed matters as the use of American tugs when berthing Vanguard Class submarines, design modifications to the Trident system, and the logistics of supplying spare parts. Walt and I had to concoct the joint communiqué at the end, a routine piece of staff work but pleasing to find that I could handle it on equal terms or better.

Back in Whitehall, my main task was to produce the Trident Project Management Plan, a sub-set of the new MOD Procurement Executive Plan, in which I was incensed to discover that the editor was proposing to have as its front cover a photograph of the new Procurement office complex being built at Abbey Wood near Bristol. I went for the jugular.

From: ADPS(SS) (me)
To: DP Plans 1a (editor of the Plan)
23 November 94

PE MANAGEMENT PLAN 95/96 – COVER ILLUSTRATION
You may wish to consider relegating the building site photograph to the sub-section dealing with infrastructure and replacing it on the front cover with the enclosed photograph of the launch of a Trident missile, lit with the first match by HMS Vanguard.

Seven days later, the Assistant Under Secretary of State for Procurement responded.

From: AUS(BS)PE
To: DP Plans 1a

I have seen Captain Thompson's minute of 23rd November. I entirely agree with him that the delivery of Trident to time, cost (with a little help from the gnomes) and quality is an outstanding achievement, of which the PE should be proud. I am therefore happy to support his proposal that CSSE's behemoth should feature on the front cover of the 1995/96 PE Plan.

Hooray! I had scored a goal. On the file copy of the correspondence, D(F&S) SS&Nuc (who he?) had scribbled: *WPB*. I have still to decode that. Google suggests: War Production Board, Weekly Playboy, West Palm Beach, Workers Party of Belgium or Waste Paper Basket. My own suggestion is Well Played Both.

This trivial bureaucratic triumph had an unexpected consequence. The Controller of the Navy, Vice Admiral (later Sir) Robert Walmsley, sent for me. He was the Lord's Anointed of Naval Engineers, the Cambridge-educated Electrical Officer I had first met in HMS *Churchill* during torpedo trials. After a meteoric career, he was now head of Naval Procurement, CSSE's boss, and a Navy Board member. CSSE's secretary delivered my summons.

'Why on earth does the Controller want to see me?' I asked.

'Don't know. I think you are about to be ennobled.'

'Ha. Ha.'

When I was ushered into The Controller's office, to my utter astonishment, he showered me with praise. His boss, the Chief of Defence Procurement, one Malcolm Mackintosh, an Australian Civil Servant who had been recruited as an outsider to pep up British Defence Procurement, had been so impressed by my management plan that he had chosen it as the model for the whole Procurement Executive. Apparently, it was exactly what he had wanted. The Controller was clearly basking in the reflected glory. I looked suitably overcome with modesty but inwardly had one enormous smile of satisfaction. The Controller had been the Director who had blackballed my appointment to his Directorate on the grounds that I may cause trouble – and I knew that he didn't know that I knew.

If you want to please the boss, first please his boss.

****

One evening Kate called me from home. 'I think I have some bad news,' she said.

Immediately I thought of the car, house collapse, big bills etc.

'What?'

'I've been diagnosed with breast cancer.'

My world collapsed.

That was it. The country would have to do without me. I would leave the Navy immediately and give her my full-time support.

The following morning I called on Richard Irwin and told him that I wished to resign with immediate effect. He then did the good Divisional Officer act on me, the first time I had been on the receiving end. The outcome was that I would be re-appointed to Rosyth Dockyard as Chief Staff Officer (Nuclear) on the staff of the Flag Officer Scotland, Northern England and Northern Ireland (FOSNNI). Rosyth was only ninety minutes from home..

CHAPTER 25

The Peace Dividend

'Peace is the only battle worth waging.'

ALBERT CAMUS

Rosyth dockyard was fighting for its survival. The Naval Base, HMS *Caledonia* (the barracks), the Joint Maritime Headquarters and several other smaller Naval establishments were already doomed. The Navy was pulling out. The Peace Dividend was biting.

The dockyard had been developed before the First World War to match the threat from the German High Seas Fleet and was one of the largest industrial sites in Scotland. It lay in the shadow of the iconic Forth Railway Bridge, its main basin being large enough for twenty-two battleships. In 1984, it had been chosen as the 'sole location' for refitting our nuclear submarine flotilla, £100M pounds having already been literally sunk in excavating a massive new docking facility for the Trident boats. The dockyard was also in the business of decommissioning old nuclear submarines and storing their de-nuclear-fuelled hulks in the now largely empty basin that was fast becoming a nuclear graveyard. Until recently, the dockyard's future had seemed secure but the naval landscape was changing fast.

During the Cold War, the nuclear Navy had been built up while the rest of the Navy was shrinking. Now, even the Submarine Flotilla was shrinking. The Navy was down to three dockyards – Portsmouth, Devonport, and Rosyth – and there was not enough work for them all. Nuclear submarine refits were where the money lay, each refit lasting almost two years and being both manpower and skill intensive. They were the survival ticket for a dockyard and only Devonport and Rosyth were in the nuclear business.

Two other new dimensions had also entered the planning calculations. First, the dockyards had just been privatised under Tory Government policy and were now in competition with each other commercially. Second, the Tory Government had also announced that Trident refits would be switched to Devonport. They stood to gain nothing politically from protecting Rosyth, which was surrounded by safe Labour seats. That had all but sounded the death knell for Rosyth.

From a purely nuclear engineering point of view, the decision to move Trident refits to Devonport was both illogical and technically ill-advised with the geology of Devonport being far less suitable than Rosyth for nuclear work. Devonport was also hopeless from a Naval personnel point of view. Trident submarines would operate from Faslane but have to go to the other end of Britain, five hundred miles away, for their refits. Where were the submariners' families to live? Where should they educate their children? They would be switching between very different Scottish and English education systems. This was not conducive to retaining highly skilled submariners.

Ostensibly, the decision was based on the dubious commercial argument that Devonport had submitted a cheaper bid for the work, but the subsequent cost of converting Devonport to meet nuclear safety standards plus the £100 million already spent at Rosyth made the marginal difference in bids insignificant. Rosyth had been robbed. It was as simple as that and it did nothing to endear the Scots to the Trident project. Twenty years later, a rampant Scottish Nationalist Party holding all but three Scottish parliamentary seats had removal of Trident from Scotland as a flagship policy.

****

Whilst the black cloud of unemployment hung over Rosyth, the black cloud of cancer hanging over Camis Eskan had lightened. Kate's diagnosis had not changed but my horrors of imminent bereavement had reduced. There would be no surgery or chemotherapy – she was to be part of a control group testing a new wonder drug and I had complete confidence that her cancer would be cured. It was simply a case of mind over matter. I was sure of that. Kate was indestructible. 'I'm fine,' was her reply to all queries about her health. She redoubled her activities with the Guides, led the Scottish contingent to an international jamboree, supercharged her Sunday school class, and attacked the garden like there was no tomorrow. 'Think positive' was engraved on her soul.

We had entered a personal phoney war.

****

FOSNNI, Admiral Sir Charles Christopher Morgan, was the most senior Vice Admiral in the Navy but no one could have been less pompous or more approachable. He had commanded the frigate *Eskimo,* been Captain of a destroyer squadron and had seen action in Brunei, Kuwait, the Indonesian Confrontation, the Beira patrols off East Africa, and the Icelandic Cod War. We were from different navies.

His previous appointment had been in Whitehall as Naval Secretary, which is the role that selects and proposes all senior officer promotions and appointments to the First Sea Lord. This included redundancies. For such a sensitive task, he had to have charm and he certainly did. He seemed able to relate to people with consummate ease yet, in a contemplative moment, confessed to me: 'People think that being an Appointer is easy but I found it extremely difficult.' It was spoken from the heart and was a measure of the man. His headquarters, a large underground bunker at Pitreavie that had played a key role during the Cold War and had been the alternative to Northwood in time of nuclear war, was to be closed.

The Navy was now engaged in re-arranging the deck chairs while the Defence budget sank. We were into New Management Strategy, Next Steps, Options for Change, a new budgetary structure, Total Quality Management, Investment in People, mission statements, management plans, blah, blah. 'This new management thing doesn't turn me on,' the Admiral proclaimed on returning from one briefing in Whitehall. He was a graduate of the Nelson school of management. He believed in caring for his people, giving his Captains clear and simple instructions, then trusting them to deliver. Subtlety was his way. As the headquarters was falling into disrepair, he made his dissatisfaction known by hosting a delegation of Ministry bean-counters in its most dilapidated room, placing seats strategically under leaks in the roof and providing buckets to catch the drips. He made his point without a word being spoken.

FOSNNI's empire stretched from the Shetlands to the Humber and included Northern Ireland, an interesting division of Britain. In FOSNNI-land, Belfast, Liverpool, Leeds, Hull, Manchester, and Newcastle were in the same country as Glasgow, Edinburgh and Lerwick.

As Chief Staff Officer (Nuclear), I was in a slightly anomalous position. I was on FOSSNI's staff but he had no executive role in the work of the dockyard, his only nuclear responsibility being, through me, for public safety in the unlikely event of a nuclear Accident. However, the matter became academic shortly after I arrived when his Chief of Staff resigned. The Admiral then sent for me and asked in his irresistible way: 'Would you mind terribly moving along in the bed and taking over as my Chief of Staff?' He made me feel that if I did, I would be doing him an enormous favour.

At the time, I saw myself as no more than a stopgap. In retrospect, I suppose it was a vote of confidence but it felt more like being invited to dine with one's psychiatrist. From his time as Naval Secretary, he would have known exactly where I stood in the officer firmament. Now he was lifting me out of my professional comfort zone into his world. I would be swimming with the Seamen.

I found the new duties somewhat nebulous and difficult to grasp. My title sounded impressive but what exactly did a Chief of Staff do? I was accustomed to executive control, to providing leadership, making things happen – or using the red card. Now, I did not seem to be in charge of anything. There were only a handful of officers actually on the staff, all perfectly competent, and FOSNNI's operations were scattered and virtually autonomous. All I could do was visit the latter to show my face, let them know we loved them, deliver bad news on closures, and deal with their signals and correspondence back at the ranch. It did, however, open up some exciting new frontiers.

The Troubles were still raging in Northern Ireland and FOSNNI operated a Northern Ireland Squadron that ran anti-gun-running patrols. The Royal Marines also operated anti-terrorist patrols on Lough Neagh in high-speed rigid inflatables. I visited them all and was treated like royalty but felt like a parasite. They were on the front line and liable to be killed while I was a military tourist. I was doing nothing for them except finding out what they did and what their needs were, but I was also putting faces to names that would arrive on my desk in promotion reports.

Similarly, I went to sea with University Reserve Naval Units in their small training craft. That too was inspiring. The student Midshipmen were keen as mustard and reminded me of myself as a Cadet at Dartmouth, but I was old enough to be their father. Again, my main role was to meet the officers-in-charge who were career officers and write their staff reports.

I was also now privy to what senior naval doctors wrote about junior doctors: comments such as *an energetic young orthopod* (I imagined a hyperactive crustacean); *a single-handed consultant* (Nelson would have approved); *a safe pair of hands* (this was a surgeon!); *a competent plumber, bricklayer and decorator, who is doing well as a surgeon* (?). My favourite was: *Entirely suited to a career in anaesthesia* with six-out-of-ten for Power of Expression.

****

On the 15th of August, 1995, Kate and I went to Perth to represent the Royal Navy at the city's Remembrance Parade for the fiftieth anniversary of VJ Day – fifty years since the dropping of the first atomic bombs. This was a civic parade with the salute being taken by the lady Provost from a saluting dais in the High Street. Veterans of the war in Asia formed the leading squad and marched past led by the pipe band of the Highland Brigade. These men had actually fought in the jungles of Burma and Malaya and some would have suffered the horrors of Japanese prisoner of war camps. For them that day, memories of their fallen comrades would have been utterly real. The stirring sound of the pipes and drums reverberating between the buildings and the sight of those elderly men marching past with such pride brought a lump to my throat. As they passed and turned their heads in salute, I knew exactly which ones were ex-Navy – they all looked me straight in the eye. I was their man. It was so very humbling.

On the saluting dais, behind the Provost, stood the Lord Lieutenant of Perthshire, a retired General. As representative of the Senior Service, I stood on the right of the line behind him with the Brigadier of the Highland Brigade on my left and a Wing Commander, Royal Air Force, on his left. At the left end of our line was a white-haired civilian gentleman with a handlebar moustache and a Victoria Cross pinned to his left breast. It was the first time I had been so close to a living VC holder. After the parade, I asked the Wing Commander if she knew who the old man was.

'That,' she said, 'is Squadron Leader Bill Reid.'

I took a deep breath.

Over coffee in the Town Hall, I approached him with due reverence and asked if he was the same Bill Reid who had attended Coatbridge High School before the war. When he said he was, I told him that I had won his Leadership Prize in 1961, just before I joined the Navy. We were both overcome with emotion. Fifty years after the Second World War had ended, he had met one of his Leadership Prize winners in the uniform of a Naval Captain and I had just met someone who had been resurrected from the dead. I had always assumed that his VC was posthumous.

Bill had been a veterinarian before the War and a bomber pilot during it. On a bombing raid over Germany, his plane had been badly damaged and two of his crew were killed. He had sustained serious wounds himself but somehow had managed to fly the crippled aircraft home single-handedly, crash-landing it on the first British airfield he found.

****

The Peace Dividend was agony for the Armed Services. I had to close Rosyth Naval Base and preside over the decommissioning of our underground bunker. Even more unpleasantly, I had to confront the Captain of *Cochrane*, the naval barracks, which also had to be closed. He had been senior to me at Dartmouth and fought tooth and nail against me. I felt unclean. Hundreds of loyal, long-serving, civilian employees would lose their jobs. It may have been the real world but it was damned unpleasant. It felt like I was sweeping up the broken pieces of a dead Navy.

One day, Chris Morgan sent for me. 'Eric,' he said, 'I'm moving my headquarters to Faslane. Would you care to join me there as the Commodore?'

Had my ears deceived me?

CHAPTER 26

'Commodore Eric'

'My Faith is in service.'

CATRIONA THOMPSON

'What are you doing now?' asked Brian Philips at my Dartmouth term's thirty-fifth reunion. He had left the Navy as a young officer to set up his own boat-building company and was now a successful businessman. We were standing on the poop deck of HMS *Victory*, once Nelson's flagship and now that of fellow Cadet, Mike Boyce, currently Commander-in-Chief of Naval Home Command and host for the reunion.

'I'm Commodore at Faslane,' I answered, feeling mildly proud of myself.

'You're a Commodore!' he exclaimed. 'You used to be a nobody.'

Being a bit slow on the finer nuances of English innuendo, I wasn't sure if he meant that I had gone up in the world or the standard of Commodores had come down. It just goes to show that if one always does one's best and keeps one's nose to the grindstone, the Navy will eventually mistake you for someone else and promote you.

I was now a Commodore. The news of my promotion was greeted with incredulity and delight in the Submarine Service, judging by the avalanche of congratulatory cards and letters I received, most expressing both astonishment and satisfaction. I would now be known as 'Commodore Eric,' a military method of bridging the formality gap when referring to senior officers. I loved that title.

This would be my final appointment in the Navy, my grand finale, and better still, I would be in uniform. Faslane was where I had spent most of my career and I knew the Base like the back of my hand. Little could I have guessed when my father drove me down to the Gareloch as a small boy to view the Reserve Fleet fifty years before that I would return as the man in charge of the country's principal nuclear submarine base.

The first requirement in any shore appointment is finding somewhere to live. In this appointment, there was the delicate matter of official residences. FOSNNI had bagged the outgoing Commodore's residence and there were Captains incumbent in all the other residences. So where was I to reside? The Captain of *Neptune* and his family were installed in the Commodore's original residence, a large house in upper Helensburgh that was to be re-assigned to me.

'Shall we evict *Neptune*?' I was asked.

'Certainly not,' I declared. 'I shall live in my own house.'

That suited me. I would not have to pay rent and there was much work to be done in the cottages but I had not factored in that I would now have a retinue. I now had a Petty Officer Steward, Leading Cook, and a driver as my personal staff. I didn't really need any of them having managed my first thirty-five years in the Navy unassisted, but neither could I cancel their posts, as my successor would need them in the Commodore's official residence.

Thus at 07.30 every morning, a staff of three reported to my semi-derelict cottage. One would make me a slice of toast, one would serve it, and the third would drive me to work. And when I travelled overnight to meetings, Simon Heesom, my Petty Officer Steward, packed my bag. That was his job. It was what he was trained to do but I found it disconcerting to be setting out without having checked the contents of my own bag. How many pairs of underpants had he allowed for a weekend away?

Bless her, Kate accepted these invaders like a mother hen with a new brood of chicks. She even tutored them for SVQs (Scottish Vocational Qualifications). Bless them, they were ever respectful of our privacy and we had more than a few laughs. At 7.30 one evening, Nobby, my Leading Cook, faithfully turned up and began to make breakfast. When I asked him why, it turned out that he had been on a run ashore in Glasgow the night before and on regaining consciousness, had mistaken 7.30 pm for 7.30 am, an easy mistake at Faslane in winter being pitch dark at both 7.30s.

It was the same Nobby who, when standing in for Simon Heesom, approached me solemnly at breakfast and said: 'Don't forget, you're presenting a quiche at ten o'clock.' That came as a surprise. Retirement gifts were usually more permanent. On checking my Commodore's Daily Programme, which Anne, my PA, lovingly typed every day, I found that I was in fact to present a quaich, the traditional Gaelic drinking cup. Nobby had never heard of that.

The next pleasant surprise was that John Tolhurst, who had joined the Navy with me, arrived as the new FOSNNI. I had not seen him since we were Cadets at Dartmouth. He had disappeared into the Fleet Air Arm as a fast jet pilot and had just been In Command of the aircraft carrier HMS *Invincible*. It proved an excellent pairing. He provided the beaming geniality and smooth sophistication of a career jet-jockey whilst I dealt with the grubby business of running a nuclear submarine base. I found it very satisfying to think that we had both committed our lives to the Navy as teenagers thirty-seven years before and were finishing our careers together. We could not have predicted that when we marched up the hill to Dartmouth on Day One.

The traditional Naval method of indicating the presence of an Admiral in a ship or shore establishment is to fly his flag from the mainmast. It is a square St George's Cross with the rank of the Admiral being indicated by the number of red balls in its white quarters. I took great pleasure in pointing out to John that I could hang both his balls from the mainmast – a Rear Admiral has two. (A Vice Admiral has one but a full Admiral has no balls at all). As baby of the flag ranks, a Commodore flies only a pennant.

Not having seen John since Dartmouth, I was unfamiliar with his behaviour in adult life and was concerned therefore to receive the following memo from his office soon after he joined.

From: Flag Lieutenant
To: Commodore

The Admiral was wondering how we are going to play the ladies when Mrs Tolhurst is away. If Mrs Thompson is unavailable, would one of the other Senior Officers' wives be willing to stand in for her?

I had not been aware that an Admiral had the right to borrow my wife! (The memo was actually about hostess duties for visiting VIPs).

****

I soon decided that having a Leading Seaman as a driver was a waste of a good sailor and so civilianised the post and returned the young man to the Fleet. A civilian driver would be perfectly satisfactory and half the cost. Thus, Mr Phil McCusker entered my life. When he arrived for an informal interview, I was shocked. He was an emaciated, forty-something Glaswegian of Irish Catholic stock, over six feet tall and with ultra short-cropped hair. He looked like an extra from *Trainspotting*, the landmark film about drug addicts in Edinburgh. Was this the image I wished to portray? At least he was dressed correctly in a smart grey suit.

He turned out to be another of the great characters to adorn my life, a man of truly independent mind with a heart of gold. In his spare time, he tended to the gardens at the Carmelite convent in Dumbarton. Phil had been born and brought up in the slums of Glasgow's notorious Gorbals, had left school to become a butcher's boy, found his way into the Navy's supply department as a storeman, converted to being a lorry driver, and had now volunteered to be my driver.

When I say 'of independent mind,' he was the real deal. He was a blackleg. He refused to go on strike. He was not a trade union brother and had been ostracised in the transport pool. Did he care? Not a fig. Phil was his own man. He didn't believe in striking. He was the sort who would never surrender his principles and simply could not do enough for me. I swear he would have swung for me had the need arisen. I had acquired not just a driver but a bodyguard.

Phil would do anything for me, which sometimes led to a spot of bother. When I happened to mention that I liked liquorice, he appeared next morning with a five-kilo bag of the stuff procured from a cash-and-carry. My bowels almost abandoned ship. When he heard Kate tell me to turn the television down, he appeared next day with headphones and a radio link for the hard-of-hearing. When the Chief of the Defence Staff (CDS), now an RAF officer, was visiting, Phil was driving us over the hill from Faslane to Coulport. 'This used to be the best place to view the Base,' I said, as we drove at speed through the forested western slopes of the Gareloch, 'but the trees now block the view.' There was a squeal of burning rubber as Phil slammed on the brakes, did a handbrake turn and headed over the edge of the road and down into the forest on a steep unmade track, stopping at a clearing a hundred metres below where felling had taken place. CDS looked utterly shocked. I was speechless. The CDS personal protection unit travelling in the car behind must have had kittens. Disappearing into the forest was not on the programme. They must have thought their charge was being hi-jacked.

'You get the best view from here, Sir,' said Phil, pulling on the handbrake.

****

Being Commodore was completely different from all my previous jobs. I was now Managing Director of an operation employing some four thousand people, not counting two thousand sea-goers, and my business was dealing with nuclear submarines, some loaded with nuclear weapons. It also included a heavy load of court and social duties, complete with ceremonial activities. I had to host an endless procession of VIP visitors ranging from the Prime Minister through to foreign Admirals. I was now lord of all I surveyed and a high visibility local VIP. Sadly that meant farewell to overalls and practical joking.

Early in my time, Captain James Burnell-Nugent (later Admiral Sir James Burnell-Nugent KCB, CBE, ADC) arrived in the Clyde in command of the frigate HMS *Brilliant*. As an old acquaintance from his submarine days, he invited me on board for a day at sea, for which he sent in his helicopter. How exciting, I was now receiving VIP treatment myself.

'Don't you miss having a depth gauge?' I quipped when I arrived in his cabin. (He had previously commanded two submarines).

'Yes, I do,' he replied in all seriousness, 'but we spend most of our time on the surface.' (?)

Brilliant was brilliant. It was my first experience of a mixed sex crew and James provided me with a delightful, young, female Leading Seaman as my personal guide for the day. In best naval tradition, she had a pigtail (untarred) pinned up over the top of her head in a compromise between military hairstyle and femininity. This young woman really knew her ship and I was most impressed but not surprised. What did surprise me was the female bunk space. It was unlike any other Naval bunk space I had seen. The girls all had colourful quilts, teddy bears, dressing gowns, and fluffy slippers. The Navy-issue sleeping bags and the austerity of a submarine bunk space were nowhere to be seen. At the end of my tour, I was delighted to find that my young guide could join me in the Wardroom for coffee. A sailor having coffee in the Wardroom with a Commodore – that was revolutionary. This was Britain's New Navy.

James Burnell-Nugent was promoted to Rear Admiral shortly afterwards and returned to the Clyde flying his flag as Flag Officer Sea Training in the aircraft carrier *Invincible*. This time, he invited me to join him off Gourock and sail up to Faslane with him. It was the first time that an aircraft carrier had visited the Base and we were all very excited about it.

On this occasion, I joined *Invincible* from the Queen's Harbourmaster's barge and had to step on to a Bermuda ladder lowered down from a watertight door several metres above the carrier's waterline. From sea level, her enormous, grey, slab-sides positively towered over me. She was underway and making six knots, not a high speed but when stepping from a small boat on to a flimsy wooden ladder sandwiched between it and the great bulk of an aircraft carrier, and with the water slapping noisily through the narrow gap between us, it felt like a pretty dynamic situation. Just as I made my death-defying leap, I heard the shrill whistle of a bosun's call from the top of the ladder. It was for me. I was being piped aboard one of Her Majesty's aircraft carriers. How's that for an Engineer?

James B-N, as he was known, went on to become Second Sea Lord, the Head of Naval Personnel, and claims the distinction of being the man who banned girlie pin-ups in HM ships and shore establishments as they were offensive to female crewmembers.

One of my early VIP visitors was the Right Reverend John McIndoe, Moderator of the General Assembly of the Church of Scotland, and his wife. As the Church of Scotland was opposed to nuclear weapons, the Moderator did not wish to be seen visiting a nuclear submarine but was keen to meet members of his Church serving at Faslane. As his interest was pastoral care, I decided that he should view an ordinary sailor's living accommodation.

In the barracks, we were met by a Petty Officer who saluted smartly and led us to what I had described as a 'typical sailor's cabin'. I had taken for granted that it would be at exhibition standard and entered confidently with the Right Reverend and Mrs McIndoe close on my heels. Oh horror of horrors! Right in front of me on the wall facing the door was a poster-size, colour photo of a naked young woman lying in a large dog basket with her legs splayed wide apart. How indescribably embarrassing. If only James B-N had made it to Second Sea Lord sooner, my blushes would have been spared.

'Don't worry, Commodore,' the good Moderator whispered over my shoulder. 'I'm a man of the world.' (Mrs McIndoe refrained from comment).

****

My own parish was large and complex. It included the main logistic support base made up of jetties, workshops, floating dock, shiplift, missile handling facilities, stores and transport, warehouses, tugs and support vessels, operations room, and communications centre. Also HMS *Neptune,* which provided Naval personnel services, accommodation, security, military police, medical, educational, and recreational facilities for 3,500 servicemen plus married quarters for 800 families. The armament depot at Coulport too, where the Polaris and Trident missiles and their nuclear warheads were processed along with Tigerfish torpedoes. I also covered the NATO armament depot up in the hills of Glen Douglas, HMS *Gannet* (the anti-submarine helicopter base at Prestwick airport), and had a Civil Service management organisation including personnel, finance and contracts departments, a public relations office, a battalion of Ministry of Defence policemen, forty-five Alsatian guard dogs, and two of Her Majesty's ferrets.

Uniquely, Faslane was still expanding against the backdrop of rapidly shrinking Defence forces. Under 'New Management' arrangements, it was now defined as an 'Intermediate Level Budget' in the new Naval Base and Supply Agency and my title had changed to Director Naval Base Clyde.

The Base Board agenda was pretty standard management stuff: budget cuts, security, property management, personnel and trade union issues, VIP visits, and nuclear safety. Inevitably, tensions rose over budget cuts with all members batting in their own best interests. This left me with some difficult decisions but with such dedicated people, chairing the Board was a pleasure. The new boy on the block was the role I had created in my previous incarnation, Director of Safety, who now wielded the red card. As the operational Squadrons were my customers, I liaised with them separately. That too was a pleasure.

****

By now, Phil, my driver, was treating me like a king. Of his own volition, he took himself on a Royalty Protection driving course at the Metropolitan Police College in Hendon. It was completely over the top for me but it did come in useful when I had to represent the Navy at an Order of the Thistle investiture in St Giles Cathedral in Edinburgh.

The Order of the Thistle is the highest order of chivalry in Scotland and equates to the Order of the Garter in England. The investiture was Scotland's greatest heraldic event of the year with Her Majesty the Queen and Their Royal Highnesses the Princes Philip and Charles and the Princess Royal plus the top tier of Scottish aristocracy all in attendance. I was representing the Navy alongside the General Officer Commanding (Scotland) and the Air Officer Commanding (Scotland). As representative of the Senior Service, it fell to me to declare the correct order of dress for the military representatives and I decreed that swords should not be worn. I had always understood that swords should not be taken into churches.

Kate and I were the first of the military reps to arrive and sat at the right hand end of our allocated pew, next to the aisle. The Air Commodore, his wife, plus an aide-de-camp arrived next followed by the General, his wife, and his aide-de-camp. I didn't mind the Air Force squeezing past but took exception to the Army; the General was wearing sword and spurs. Having a man with sword and spurs squeezing past in the narrow confines of a Presbyterian Church pew scared the hell out of me and I feared for Kate's nylons. Why he needed spurs when travelling in a staff car defeats me. He had as much need for them as I had for a Geiger counter. When he had installed himself, he leant across Kate as if she did not exist and spoke to me.

'Don't you have an aide-de-camp?' he asked, as if I had forgotten my trousers.

'No,' I replied. 'I can find St Giles on my own.'

At the end of the service, we filed out of the cathedral just as the olive-green uniformed Royal Company of Archers, the Sovereign's Bodyguard in Scotland, were marching off from their position in front of the cathedral steps. Right behind them, Phil arrived in my British racing green Rover staff car. He leapt out and opened the rear door for Kate and me to climb in. We were the first and only guests to be picked up at that point, with no other vehicles being allowed into Cathedral Square. Having completed the Royalty Protection course, Phil had known all the right code words to pass through the police security barriers.

And so we left the General and Air Commodore stranded on the steps of St. Giles with their noses well out of joint. 'Go and find our bloody staff car,' I heard the Air Commodore growl angrily to his young aide-de-camp as Kate and I drove off.

One up to the Navy!

****

Dealing with trade unions is probably not the sort of activity most members of the public would associate with naval officers but I found myself responsible for exactly that. I had to chair the Base union-management meetings. These were usually amicable and cooperative affairs as significant issues such as Civil Service pay and conditions were decided at national level, but one policy change from on high led to an extraordinary development. In 1997, the New Labour Government of Tony Blair swept into power with a landslide majority and shocked the trade union movement by pursuing the previous Tory government's policy of privatisation. The Base was to be privatised.

The unions vehemently opposed this and so did I – against the Ministry line – not because I supported the unions or was anti-privatisation but because I could see huge efficiency savings that an incoming commercial operator would implement immediately and take as profit at the taxpayers' expense. I had identified such savings – they were obvious – but I did not have the authority to implement them. For example, there were two main workshops in the Base, one belonging to the Civil Service and one to the Navy, which a private operator would merge immediately. Had the Ministry been more commercially astute, it would have invoked these changes before privatisation. Turning public assets into private profit was hardly a principle of the Labour Party.

The threat of privatisation led rapidly to a confrontation in Whitehall when the unions demanded a meeting with the Minister for the Armed Forces, Dr John Reid MP, (now Lord Reid). I attended as one of his advisers. Sitting opposite him were: Jack Dromey, leader of the Civil Service trade unions (husband of Labour Minister, Harriet Harman, and now a Labour MP himself), John McFall, our local Labour MP (also now a Lord), plus Jim Flynn and Alan Grey, respectively the chief shop stewards in my Industrial and Non-Industrial trade union committees. The meeting was in effect an internal Labour Party punch-up and was opened spontaneously by Jack Dromey before the Minister had a chance to speak.

'This is not how a Labour Government behaves, Minister,' he declared, staring straight at John Reid and tapping the table hard with his index finger. Clearly, Dromey was of the trade union school that believed it was for the unions to tell Labour Governments what to do. He was physically a much larger man than I had expected and lacked nothing in self-assurance.

John McFall then sprang his ambush. 'You can't proceed with this privatisation, Minister. This is what the Party said in its election leaflets.' He produced a Labour Party leaflet from his briefcase and laid it before the Minister. It promised that Labour would cancel Conservative plans to privatise the Base.

The Minister was snookered but John Reid was a seasoned campaigner. 'Well, gentlemen,' he said, 'let's follow normal negotiating procedure. Would you please withdraw your delegation while I confer with my colleagues?'

'How much will this privatisation save us?' he asked me when they had left the room.

'£450k per annum.'

'Well, what's a £150K between friends? Call them back in.'

The union delegation returned.

'Right, gentlemen,' he announced, 'I shall call off this privatisation on the condition that you can find £300,000 of savings. I'm giving you six weeks to do it.' No doubt he thought there was no chance of that.

I gasped. I managed the Base, not the unions. They could not do anything without management support. The ball was now in my court.

As we left the meeting, I tapped my shop stewards on the shoulders and invited them for a pint at the Red Lion in Whitehall. We needed to hatch a plan. They needed my support and I needed their cooperation – we were in this together. As Sir Francis Drake famously said: '*Let us be all of one company.*' That was the ethos of the Royal Navy but an alien concept in the trade union-management world.

Over an amicable pint of best bitter, we agreed that I would take two senior managers and the two chief shop stewards out of the line to form a joint attack team with the remit to find £300,000 of savings in six weeks. They did it in six days. Trade unionists know better than anyone where shop floor inefficiencies are to be found.

To the chagrin of my masters in the Ministry, we had succeeded. The privatisation was called off. It did me no favours in the corridors of power but I was proud of our achievement. It was management at its best, but not without pain. It cost Jim Flynn, chief shop steward of the Transport and General Workers Union, his job, not through consequential redundancy but through the abuse he received from his fellow workers. He quit.

A few days after he left, I received an invitation to his farewell party at a bowling club in Dumbarton. I counted that as a great honour. Not many managers let alone senior military officers receive invites to a shop steward's party.

****

Since the arrival of Polaris submarines, Faslane had been one of the most politically sensitive and highest public profile military establishments in British Defence, along with Greenham Common when US Intermediate Range nuclear weapons were based there. Both locations had acquired 'peace camps' at their gates, the campers' purpose being to create a constant stream of adverse publicity and thus draw public attention to their anti-nuclear propaganda.

At Faslane, the camp was a garish, ramshackle collection of dilapidated caravans, an old bus, and an assortment of shanties populated by a mixed bag of protesters. They ranged from nuclear disarmament campaigners, pacifists, environmentalists, conscientious objectors, Christians, and anarchists all seeking to bring an end to our Nuclear Deterrent. Some, I know, were utterly sincere. I befriended two who lived locally and we continue to share mutual respect but agree to disagree. Others, I suspect, were simply anti-establishment or dropouts who would have joined any protest regardless of subject. Some were little more than vandals, spraying obscene graffiti and damaging equipment in Ministry outstations. In the Base, we regarded ourselves as the true peace camp; theirs was the protest camp.

Greenham Common had a different flavour. It was driven by a feminist agenda and was largely women only, the principle being that women represented motherhood and therefore the future of our children.

As a member of the Armed Forces, I was not allowed to engage in political activities, but boy did I want to remind the protesters that the country had been at peace since 1945 thanks to our Nuclear Deterrent. Our war-weary leaders who had been through the hell of the Second World War had not introduced Nuclear Deterrence to start a Third World War but because they never wanted to see another one.

The protesters completely failed to recognise that it was the military, not pacifists, which defeated Fascism. The military had liberated occupied countries like France and Holland, and then faced down the Soviet's ambition to destroy our freedom. It was the threat of Mutually Assured Destruction, not pacifism, which had ensured the Cold War did not descend into the Third World War. The Cuban missile crisis is proof of that.

It is astonishing to reflect that the Second World War began a mere twenty-one years after the end of the First World War, a war so terrible that it had been dubbed *The War to End All Wars*. The question is whether mankind will do it for a third time? The post Second World War belief was that nuclear weapons were the only thing that could prevent it. That has been proved correct. Forty-six years of Cold War had ended without it descending into the Third World War. The proof of the pudding is in the eating.

Had we adopted a non-nuclear policy at the end of the Second World War, the risk of world war during my lifetime would have been massively increased, as would the number of troops we required. Of that, I have no doubt. We could have had a Vietnam in Europe. The Soviet Union was openly committed to establishing worldwide Communism. It had not hesitated to invade Poland, Hungary, and Czechoslovakia when these countries attempted to establish free, independent democracies. Had Western Europe not had the benefit of NATO with its nuclear counter-strike capability, the Soviet Union would have been much more aggressive in seeking to bring all of Europe under its influence. And had the Soviets ever achieved their cherished aim of splitting the NATO Alliance and separating Western Europe from its American allies, they would still have had to reckon with independent British and French nuclear counter-strike capabilities.

I did have some sympathy for the protesters. They were idealists, the foot soldiers of protest, the poor bloody infantry. They lived in squalor on the verge of a main road motivated by the belief that nuclear weapons are evil. I agree. It is their sheer horror that makes them a deterrent. I don't think blowing kisses would have impressed the Kremlin. I too fervently believe in peace and am proud to call myself a pacifist but I was an active pacifist. In my own small way, I had been taking positive steps to ensure peace and not leave it to chance.

In arguing against nuclear deterrence, one must never lose sight of the devastating human cost of the Second World War, which was slaughter on a global scale. More than sixty million people lost their lives, less than half of one per cent of them from nuclear weapons. Mankind does not require nuclear weapons to create megadeath. The rifle has killed far more people than nuclear weapons – and so has the motorcar.

By the end of the Second World War, the league table in millions-of-dead stood as follows:

Country	Military dead	Civilian dead	
China	1.5	20	
USSR	13.6	5	
Poland	0.6	6	
Germany	3.5	2	
European Jews	–	6	Exterminated by Nazis
Japan	2.6	0.95	
France	0.245	0.173	Surrendered/Occupied
United Kingdom	0.42	0.07	
USA	0.29	0.00001	
Total dead	**22.755**	**40.193001**	

(Source:Readers Digest History of World War Two)

These simplified figures show that invaded countries suffered most, namely China, USSR, and Poland. They also suffered vastly more civilian deaths than their German and Japanese aggressors. (The Germans actually exterminated more of their own Jewish citizens than they lost as casualties of the war they had caused). And without having attacked anyone, Poland was invaded twice, once from either side, and suffered most per head of population. The figures also show that Britain and America who were not invaded, suffered least, which leads to the obvious conclusion that one should never allow one's country to be invaded.

The end of the Cold War had now delivered a tangible Peace Dividend. In 1994, the Prime Minister announced that British strategic missiles had been de-targeted, and that the number of nuclear warheads in our Trident submarines had been reduced to a maximum of forty-eight and that the notice to fire had been extended to several days. Our submarine flotilla was also to be reduced to fourteen submarines (ten SSNs and four SSBNs), whereas at the height of the Cold War we had twenty-eight and the Soviets four hundred.

On a more human scale, a Russian (not Soviet) nuclear submarine on patrol to the West of the Hebrides – probably searching for *Victorious,* our on-patrol Trident submarine – sought British help for the first time in rescuing a young Russian submariner with acute appendicitis. We collected him by helicopter and delivered him to hospital in Stornoway. Under the Soviet regime they would have let him die. A British submarine had also now officially visited the ex-Soviet submarine base at Severomorsk and a Russian submarine would visit Faslane in 2001.

The Western World had just reached a historical zenith in terms of peace. The nuclear superpowers had finally agreed that they were not about to destroy each other. In fact, post-Soviet Russia was no longer regarded as a superpower, though this was merely an interregnum. It would take only nine years for Russia to find a new autocrat in ex-KGB officer Vladimir Putin. He was elected President in 2000 and hankered after a return to the glory days of the Greater Russia that had disintegrated at the end of the Cold War. He now seeks to show the world that Russia is still a superpower. He has refreshed his armed forces, renewed his nuclear arsenal, invaded the Crimea (Ukraine's sovereign territory), and currently (2017) has deployed his air force to attack rebels in Syria, bombarding them with cruise missiles from his nuclear submarines. Kipling's Great Game had only been on pause. Putin is not yet the threat we faced during the Cold War but peace should never be taken for granted.

Philosophically speaking, having a protest camp at the gates of the Base was a healthy sign. It meant that freedom was alive and well in the United Kingdom. This was what my father and so many peace-loving men of his generation had fought to defend against the tyranny of Fascism. It was what the Nuclear Deterrent was protecting. I did not agree with the protesters' view but I would defend their right to peaceful protest. Ultimately, the electorate decides.

However, the protesters' ability to fire up the media with spurious shock-horror stories could be a pain-in-the-neck. On one occasion, they fed an entirely specious report to the press about a major nuclear drama unfolding in the Base involving helicopters landing on the hillsides, Strathclyde Police appearing in numbers, and the Secretary of State for Defence (George Robertson MP) arriving in secret. It was a classic case of two and two make five. The three activities were entirely unrelated: the Minister was on a routine visit, the police were preparing for an anticipated anti-nuclear demonstration, and the helicopters were part of a Royal Marine exercise on the local Army training area. Nevertheless, this non-event made front-page headlines. My Public Relations department could never relax.

Baron Robertson of Port Ellen, as he is now known, was no stranger to protest. He told me proudly that he had been arrested as a teenager in Dunoon for participating in an anti-nuclear protest against the basing of American Polaris submarines in the Holy Loch. He was now the Labour Party's Minister of Defence with ultimate responsibility for Britain's Nuclear Deterrent and would go on to become Secretary General of NATO, an alliance backed by British and American nuclear weapons. I never did discover when he abandoned his anti-nuclear principles.

From the Base point-of-view, the 'peace camp' was scarcely a sideshow, but it had the benefit of keeping our security force on its toes, the latter's main concern being an IRA terrorist attack. However, the campers managed to make us look stupid on several occasions.

One Christmas (before my time as Commodore), three peace-campers dressed as Santa Clause not only infiltrated the Base but actually boarded *Repulse,* a Polaris submarine, berthed in our most secure area. They had breached the security fence using ladders and, as luck would have it, arrived on the jetty just as the submarine's trot sentry had gone aft to do his hourly check of the stern draught marks. They were thus able to nip across her gangway unchallenged. When they reached the submarine's Control Room, they issued a CND press release by telephone before being apprehended by sentries armed with pickaxe handles.

What the intruders did not know was that whilst on the jetty, they had been in the gun sights of a Royal Marine guard. He could have shot them dead and was entitled to do so had he judged them to be about to sabotage one of the country's 'vital assets'. The Marine had only moments to decide. He held fire. Thank God. Marine Kills Santa Claus would have been an even more unpalatable headline than Protesters Board Submarine. Had that incident happened in the Soviet Union or the USA, they would have been shot. The Soviets shot innocent civilians merely for trying to escape to freedom over the Berlin Wall.

Infiltrating the Base was just a game for the protesters. They knew they would not be shot, that they would be handed over to the police, and would then be released to have another go. It may have been a good game for them but it was not a game for those charged with maintaining Base security. When the headline 'RAIDERS GET INTO N-SUB NERVE CENTRE' appeared in a national daily (*Daily Express, 11 Oct 88*), our local Labour MP (John McFall) demanded that heads should roll. One did. The Captain of *Neptune*, who was responsible for security, was sacked and his career ruined. One of my great fears was that faced with similar embarrassment, my own head would roll or I would have to initiate similar disciplinary action against another good officer who had been forced to fight with his hands tied behind his back.

When I arrived as Commodore, the 'peace camp' had been re-vitalised by three separate events: the arrival of Trident; the protesters at Greenham Common having turned their attention to Faslane after the Americans had left; and the third that the good citizens of Helensburgh had voted to secede from Labour-controlled, anti-nuclear Dumbarton District Council and join the Independent-controlled, pro-Navy Argyll and Bute Council. It then declared an intention to evict the 'peace camp' as it was annoying the local residents. Argyll and Bute's initiative lit the blue touch-paper on the international protest movement. Under the threat of eviction, protesters piled in from across the globe. Faslane was now their destination of choice. The camp positively fizzed with energetic new activists, not least an environmental protectionist called 'Swampy', fresh from a failed protest against runway extension at Manchester airport. He was a tunneler and his intention was to tunnel under our by now very sophisticated security fence in a sort of Great Escape in reverse.

Female activists from Greenham Common were now also taking the lead. In a change of tactic, one topless woman paddled into our secure area by canoe, followed by a photographer. What she wanted was a photograph of Ministry of Defence Police officers manhandling a semi-naked woman. That would have guaranteed the front pages and depicted the Base as brutal and militaristic. (The police Marine Unit kept her at bay until she became cold and returned to the beach).

Two women in rubber diving suits, one from New Zealand, then swam under our jetties at night from Timbacraft, an adjacent small shipyard. Once under the jetties, they could not be seen and it was easy for them to choose their moment to clamber on board the tail of a nuclear submarine. They made us look so damned inept. But what would the verdict have been had my security team shot them as suspected saboteurs? How were my armed guards to decide in the dark if two unauthorised frogmen in the water beside one of our submarines were terrorists or protesters?

Three other women boarded an unguarded Admiralty Scientific Service research barge moored in remote Loch Goil and smashed twenty computers, throwing logbooks, files, and other equipment overboard. This was pure vandalism and cost the taxpayer £100,000. They were arrested, tried in Greenock Sherriff Court on 21st October 1999 and, incredibly, were acquitted. Their defence was that their crime was preventing the greater crime of using nuclear weapons. No it was not, but Sheriff Margaret Gimblett agreed. She concluded the trial with the following statement:

> *'You have been found not guilty and you are therefore free to go. Yesterday I made it clear that the courts do not normally allow a crime to be committed to prevent other crimes except in very special circumstances. There were such circumstances in this particular case.'*

The judgement made my blood boil.

The juxtaposition of the Base and the 'peace camp' brought to mind Robert Burns' brilliant social satire, *The Twa Dogs*. In this, a poor shepherd's Collie and a wealthy laird's Labrador compare notes on the lifestyles of their respective masters, then: *'Rejoic'd they werena men but dogs.'*

****

Whilst I was Commodore, HMS *Repulse,* the last of the Polaris submarines, decommissioned and that signalled the end of an era. On a personal basis, it was hugely symbolic. The Bahamas Agreement had been signed when I was a Midshipman and Polaris had spanned my career almost exactly. It had been my generation's role in the Cold War, our contribution to peace and freedom. It seemed so fitting that I should be Commodore at Faslane when this hugely successful programme came to an end.

In principle, the event was a private celebration for the *Repulse* crew and their families but in attendance were the Prime Minister (John Major) and the Minister of Defence (Michael Portillo). The Naval top brass included the First Sea Lord (Sir 'Jock' Slater), Flag Officer Submarines (Sir James Perowne) and FOSNNI (John Tolhurst). It also included a posse of retired senior US Navy officers who had been involved in the signing of the Bahamas Agreement. That greatly reinforced the sense of our blood brotherhood – the Americans were so overwhelmingly on our side.

Highly symbolically, alongside the new Trident jetty lay HMS *Victorious,* the second of our Trident boats, preparing for patrol and also the USS *West Virginia,* an American Trident boat that had detached from patrol to be in attendance. In terms of the Submarine Service, it was a historic and inspiring occasion.

In addressing the assembly, the Prime Minister said: '*Throughout turbulent years, the Polaris force has always been there – always ready, always prepared, always the ultimate guarantee of this country's security. I have no doubt that we were right to maintain a minimum credible strategic nuclear deterrent for the UK. We will continue to do so for as long as our security needs require us to do so. It would be folly for us to act differently. Even though the circumstances have changed, the world still remains an uncertain and dangerous place.*'

When *Repulse* sailed for the last time, she sent this valedictory signal.

1. *Our final departure from Faslane was at once sad, memorable and moving. Very many thanks for making it such a warm and grand occasion.*
2. *We can never thank you enough for 28 years of unstinting support.*
3. *Thank you and farewell.*

Between 1968 and 1996 our four Polaris boats – *Resolution, Renown, Repulse* and *Revenge* – had completed an unbroken sequence of 229 Strategic Nuclear Deterrent patrols and had, without fail, remained continuously at fifteen minutes notice to fire – a simply astonishing achievement. Their mission had been accomplished. My steam leak emergency in *Revenge* was the closest we had come to failure. We had been one flick of a switch away from it. Polaris had been the very essence of the Cold War. As someone remarked: 'It was the best insurance policy the nation ever had.'

For twenty-eight years, somewhere out there in the deep ocean, dedicated crews of British submarines had quietly and anonymously stood guard over the nation's peace. Few British citizens would have been aware of this let alone have spared it a thought. The moment prompted me to write this poem:

PEACE BE WITH THEM

In the bowels of the beast with a heart of steel,
in Neptune's black abyss,
Stand sixteen silent sentinels on watch o'er Britain's peace.
And through the black abyssal deep,
each day of every year,
The Reaper ploughs the ocean and sows his seeds of fear.

In the bowels of the beast with a heart of steel,
where the nuclear cauldron boils,
A hundred brave submariners attend their awesome toils,
Whilst snug in quilted feather beds,
full fifty million sleep,
And spare no thought for those at sea, nor pray their souls will keep.

****

As Commodore's wife, Kate was expected to act as a hostess for visiting VIPs and as the unelected senior wife in the Base, she also chaired the Wives Club. These were not roles she would ever have sought but she was perfect for them. With her old-school training as a Home Economics teacher, she knew the rules for gracious living. With her father being a Gaelic-speaking crofter's son from Lewis, she knew about thrift and humility. With her mother being a lawyer's daughter, she knew how to behave socially. With her lifetime commitment as a Guide Leader, she knew how to rough it in camp with teenage girls and from her time as a Sunday school teacher, she was imbued with the spirit of Christian compassion. She exuded such warmth and sincerity that she could communicate with everyone, from Her Royal Highness Princess Anne to a young sailor's wife. Everyone loved her.

Kate never sought to impose herself. Her style was to listen and laugh as often as possible. Bad feelings were simply not possible in her company. Six years later, when she was admitted to the acute cancer ward in the Beatson Institute in Glasgow for the last time, one of her fellow Sunday school teachers said to me: 'Kate's Faith will get her through.' When I repeated that to Kate, she replied: 'My Faith is in service.' That summed her up to a tee. She was a giver. She sought nothing for herself. She had completely fooled me about the true state of her health and when the ghastly truth could no longer be hidden, she apologised for ruining my life.

When the time came for me to ride off into the Naval sunset, I requested that Kate should be at my side and so, having formally handed over to my successor, she and I took our seats side-by-side on an industrial trailer decorated with bunting, to be towed by tractor to the jetty where we boarded the Queen's Harbourmaster's launch. After thirty-seven years, I was departing Her Majesty's Nuclear Service in uniform with the woman who had supported me for thirty-two of them at my side. As we sailed past the array of ships and submarines alongside the jetties, they all saluted, and then we headed down the Gareloch with our Royal Marines and Ministry of Defence Police escort criss-crossing our track in their high-speed rigid-inflatables like the Red Arrows. We even had a helicopter from HMS *Gannet* overhead with a White Ensign dangling beneath. It was a sensational way to end my Naval career.

For icing on the cake, I had arranged that my mother would be waiting to greet us at Rhu Marina at the south end of the loch. Thirty-seven years after waving me off to join the Navy, she was symbolically welcoming me home. During the two World Wars, millions of mothers had never seen their sons return.

I had lived through the whole second-half of the twentieth century. Had I lived in the first half, I would have had to fight in two world wars and probably would not have survived. I had truly been blessed. I was from the luckiest generation ever to have walked this earth. I had served in the Royal Navy for thirty-seven years and never known war. I had lived my life in peace.

For thirty-seven years I had never failed to do my duty, just as I had promised as an eight-year-old Cub, and death alone would part me from Kate – that too I had promised.

CHAPTER 27

Leros

'For your tomorrow, we gave our today.'

THE KOHIMA PRAYER

As the small inter-island aircraft rounded the massive thunderhead that had delayed my departure from Athens, I caught my first sight of Leros. It must have looked exactly the same sixty-six years earlier when the German pilots lined up for their bombing run on Port Lakki. Then I had been in my mother's womb and my father was serving in HMS *Intrepid*, a British destroyer lying at anchor in Lakki Bay. They sank her.

The surprise attack was the opening shot in the short but bloody Battle of Leros in which the Germans triumphed. Because of the sinking, the name had been imprinted in my memory since childhood, but I knew nothing about the place other than that it was in the Dodecanese Islands. This was my first visit.

I was now both retired and widowed and was joining Henry Buchanan, my old Naval chum. In his retirement, he had bought a yacht and was cruising in the Greek Islands. As nuclear submarine engineers, we were members of a very special band of brothers. We had joined the Navy together, stayed together through officer training at Dartmouth, through Engineering degrees at the Royal Naval Engineering College, through submarine training, and through post-graduate nuclear training at Greenwich and Dounreay. After all that, we had spent our careers On Her Majesty's Nuclear Service. We were tried and trusted officers-and-gentlemen and I felt privileged to count myself a member of such a brotherhood.

Henry had done his homework and whisked me away immediately to the small taverna opposite the tiny airport building. The family who ran it spoke no English and we could manage only 'good afternoon' in Greek but to my amazement, I was greeted like a returning hero. Leros had been brutally savaged by the Germans in retribution for Italy surrendering the island's naval base to the Allies. The father of the patron had fought in the Greek Resistance and the family had not forgotten the simultaneous sinking of *Intrepid* and *Queen Olga*, the latter being the pride of the

Greek Navy. Before coffee could be served, the family's wartime photograph albums were laid before me in a touching show of solidarity with a son of their wartime allies.

On the wall of the entrance tunnel to the local war museum, there was a photograph of *Intrepid* and *Queen Olga* in flames prior to sinking. It was the first photograph of *Intrepid* I had ever seen, sobering to think that my father had been on board at the time, fighting for his life. *Queen Olga* had sunk rapidly and lost most of her crew, a national tragedy for the Greeks. *Intrepid* had taken twenty-four hours to capsize and sank with much smaller loss of life.

At Leros I also discovered a small Imperial War Graves Commission cemetery. In its entranceway stood a simple monument behind which lay neat rows of uniform headstones, the final resting places for the eighty-eight Allied soldiers who had fallen in the battle. In a corner stood a separate group of six headstones huddled much closer together, almost touching each other. These were the graves of my father's shipmates.

There is something profoundly moving about war cemeteries. It may be because they are the eternal resting places for comrades-in-arms who fell together. It may be the sheer number of graves and the drill order of their layout. It is certainly much to do with the age of the fallen for they were all young men. I have seen the fields of headstones for those who fell in the Normandy landings. I have driven through the acres of war graves in Northern France for the millions who fell as cannon fodder in the trenches of the First World War, and I have visited the cemetery in Berlin for the Allied airmen who were shot down. I have shed tears in them all and left with only one thought: this must never happen again. It has taken nuclear weapons to prevent it. That has been the price of peace.

The feeling inspired by these cemeteries is nowhere better expressed than in the Kohima Prayer, engraved on the war memorial at Kohima in India where, in 1944, the British and Indian armies halted the advance of the Japanese with great loss of life. This prayer has been repeated at every Remembrance Day service since the end of the Second World War. It says simply:

> When you go home, tell them of us and say,
> For your Tomorrow, we gave our today.

I have lived that 'Tomorrow'.

Glossary of Terms

ASDIC	Antisubmarine Detection Investigation Committee
AGI	Acronym for an intelligence gathering spy ship
AUWE	Admiralty Underwater Weapons Establishment
BSc(Eng)	Bachelor of Science (Engineering) degree
BUPA	British United Provident Association
CB	Citizens' Band (radio)
C-in-C	Commander-in Chief
CIA	Central Intelligence Agency
CND	Campaign for Nuclear Disarmament
CO	Commanding Officer
Dartmouth	Britannia Royal Naval College
DNA	The molecular genetic code fundamental to all living organisms
EEC	European Economic Community
EURATOM	The European Atomic Energy Authority
EW	Electronic Warfare
Fleet Tender	Small vessel used to service ships in the Royal Navy, c.f. water taxi
FOSM	Flag Officer Submarines
For'ard	the naval word for 'forward' in a ship
HF	High Frequency
HMS	Her Majesty's Ship
HQ	Headquarters
IAEA	The International Atomic Energy Authority
ICBM	Inter Continental Ballistic Missile
IQ	Intelligence Quotient, a measure of a person's true intelligence
IRA	Irish Republican Army
IRBM	Intermediate Range Ballistic Missile
KGB	Soviet secret police
TRV	Torpedo Recovery Vessel

Manadon	Royal Naval Engineering College
MENSA	an organisation for those who score over 98% in intelligence testing
MEM	Marine Engineering Mechanic
MI6	British Military Intelligence
MOD	Ministry of Defence
MOD(PE)	Ministry of Defence Procurement Executive
NATO	North Atlantic Treaty Organisation
NASA	National Aeronautical and Space Administration
Niaff	nonentity
Nipple	a connection point on an Emergency Breathing System
Oggin	Naval slang for the sea
OLQ	Officer-like Quality
PAG	(Nuclear) Procedural Authorisation Group
PE	Procurement Executive
Perisher	Submarine Commanding Officers Qualifying Course
Pipe	Traditional method of Naval signalling with a whistle known as a bosun's call
Pusser	Slang for the Navy (Pusser's rum means Navy rum)
RAF	Royal Air Force
RN	Royal Navy
RS	Radio Supervisor
SAS	Special Air Service
Scram	the term for shutting down a nuclear reactor
SEATO	South East Asia Treaty Organisation
SM2, SM3, SM10	SM refers to a submarine squadron and is used in referring to its Captain
Snort	the submarine version of snorkelling
SSN	nuclear-powered hunter-killer submarine
SSBN	nuclear-powered ballistic missile carrying submarine
SSK	diesel powered submarine
SOSUS	Sound Surveillance System, a network of top-secret, US, seabed listening arrays
STWG	Submarine Tactics and Weapons Group
SWEO3	Squadron Weapons Electrical Officer, Third Submarine Squadron
TG	Turbo generator
TRV	Torpedo Recovery Vessel
TTU	Torpedo Trials Unit
Trot sentry	upper deck sentry of a submarine
UN	United Nations

USSR	Union of Soviet Socialist Republics
VHF	Very High Frequency radio
VC	Victoria Cross
VJ	Victory over Japan
VLF	Very Low Frequency
UHF	Ultra High Frequency
WRA	Wardroom Attendant (usually a naval pensioner)
XO	Executive Officer
Yardarm Clearing	covering one's back